What Your Colleagues Are Saying . . .

"Our district has been successfully utilizing Dr. Nicki Newton's Guided Math approach for the past two years in our K–5 classrooms. By focusing on UDL and SEL, *Math Workshop Plus* provides easily applicable resources to enhance what teachers are doing in their classrooms to ensure equity, access, and excellence for *all* students."

Erin Gomez
Assistant Superintendent of Schools
Middlesex, NJ

"This book is the perfect combination of the best math practices and fresh new content. It's clear that the authors have authentic experience with the topic. They address the real struggles that teachers have implementing Math Workshop by organically including SEL, UDL, and the diversity of all the humans we work with daily."

Lori Breyfogle
Elementary Math Specialist Coach
Imperial, MO

"We all understand that everything we do in the classroom can be improved to enhance student learning. New techniques, shifts in strategies, and enhanced tasks, all make us better. This book is a treasure trove of such practical techniques and strategies. I love the core thesis here that the math workshop model can mature into *Math Workshop Plus*—an enhanced approach for helping us link the math workshop model, Universal Design for Learning, and SEL approaches in every lesson we teach."

Steven Leinwand
Consultant and Researcher
Washington, DC

"In *Math Workshop Plus*, Alison J. Mello and Dr. Nicki Newton show us exactly how to design our classrooms to make math accessible to every student, creating a culture where all kids can own their success. It's an absolute must-read for anyone looking for actionable and impactful ways to bring equity and joy to math education, irl!"

Vanessa "The Math Guru" Vakharia
Author, *Math Therapy*™
Host, the Math Therapy Podcast
Toronto, Ontario, Canada

"*Math Workshop Plus* provides relevant research and ample resources to equip mathematics educators with the knowledge and skills necessary to meet the diverse learning needs of all students within the mathematics classroom while working to ensure equitable outcomes for *all* students!"

Janet D. Nuzzie
District Intervention Specialist, K–12 Mathematics,
Pasadena Independent School District
Pasadena, TX

"*Math Workshop Plus* by Alison J. Mello and Dr. Nicki Newton masterfully integrates Universal Design for Learning and the CASEL framework with the math workshop model to elevate mathematics instruction. Centered on equity, this book provides practical, classroom-ready strategies that empower teachers and coaches to transform their math workshop to ensure equitable outcomes for every child."

Georgina Rivera
Principal, West Hartford Public Schools
Vice President (2021–2023), NCSM
West Hartford, CT

"Mello and Newton deliver once again with another amazing book to help educators get this work done! Through practical tools, resources, and tips they've elevated the Math Workshop model to be even more accessible for children by weaving in UDL and SEL to create the *Math Workshop Plus* model!"

Melanie Harding
K–3 Mathematics and Science Coordinator
Trenton, NJ

"*Math Workshop Plus* is a comprehensive guide to planning effective math instruction for the wide range of learners in today's classrooms. It's a must-have resource for teachers who believe that one size does not fit all in the math classroom and that children's mathematical thinking deserves to be seen, heard, and celebrated!"

Kate Ariemma Marin
Assistant Professor, University of Louisville
Louisville, KY

"I'm very grateful to Alison J. Mello and Dr. Nicki Newton for writing this book that provides specific and actionable ideas for meeting the various needs of all our students in ways that provide access and equity for all students to rich mathematical experiences. Throughout the book they discuss how we can be intentional in our classroom structure through the math workshop model that not only develops the mathematical understandings of our students but facilitates the building of positive math identities and agencies of all our students."

Ann Elise Record
Consultant, Ann Elise Record Consulting LLC
Concord, NH

Math Workshop Plus
Grades K–8

*I dedicate this book to the many teachers, students, and mentors
who have taught me that learning is a process and a journey that never ends.*

—Alison J. Mello

To Mom and Pops always.

—Dr. Nicki Newton

Math Workshop Plus Grades K–8

Elevating Your Practice to Ensure Personal and Academic Success for All Students

Alison J. Mello
Dr. Nicki Newton

CORWIN

FOR INFORMATION:

Corwin

A SAGE Company

2455 Teller Road

Thousand Oaks, California 91320

(800) 233-9936

www.corwin.com

SAGE Publications Ltd.

1 Oliver's Yard

55 City Road

London EC1Y 1SP

United Kingdom

SAGE Publications India Pvt. Ltd.

Unit No 323-333, Third Floor, F-Block

International Trade Tower Nehru Place

New Delhi 110 019

India

SAGE Publications Asia-Pacific Pte. Ltd.

18 Cross Street #10-10/11/12

China Square Central

Singapore 048423

Vice President and Editorial
  Director: Monica Eckman

Associate Director and
  Publisher, STEM: Erin Null

Senior Editorial Assistant: Nyle De Leon

Production Editors: Tori Mirsadjadi and
  Amy Schroller

Copy Editor: Amy Hanquist Harris

Typesetter: C&M Digitals (P) Ltd.

Proofreader: Dennis Webb

Indexer: Integra

Cover Designer: Candice Harman

Marketing Manager: Margaret O'Connor

Copyright © 2026 by Corwin Press, Inc.

All rights reserved. Except as permitted by U.S. copyright law, no part of this work may be reproduced or distributed in any form or by any means, or stored in a database or retrieval system, without permission in writing from the publisher.

When forms and sample documents appearing in this work are intended for reproduction, they will be marked as such. Reproduction of their use is authorized for educational use by educators, local school sites, and/or noncommercial or nonprofit entities that have purchased the book.

All third-party trademarks referenced or depicted herein are included solely for the purpose of illustration and are the property of their respective owners. Reference to these trademarks in no way indicates any relationship with, or endorsement by, the trademark owner.

Credit for icons that recur throughout the book is attributed as follows: SEL icon—Istock.com/glopphy; Action icon—Istock.com/Turac Novruzova; Try It icon—Istock.com/Ankomando; and Equity Check icon—Istock.com/fendy hermawan.

No AI training. Without in any way limiting the author's and publisher's exclusive rights under copyright, any use of this publication to "train" generative artificial intelligence (AI) or for other AI uses is expressly prohibited. The publisher reserves all rights to license uses of this publication for generative AI training or other AI uses.

SEL Icon Source: iStock.com/Glopphy

Action Icon Source: iStock.com/Turac Novruzova

Paperback ISBN 978-1-0719-3263-6

DISCLAIMER: This book may direct you to access third-party content via web links, QR codes, or other scannable technologies, which are provided for your reference by the author(s). Corwin makes no guarantee that such third-party content will be available for your use and encourages you to review the terms and conditions of such third-party content. Corwin takes no responsibility and assumes no liability for your use of any third-party content, nor does Corwin approve, sponsor, endorse, verify, or certify such third-party content.

Contents

Preface   xi
Acknowledgments   xiii
About the Authors   xv

Chapter 1. The Big Ideas of Math Workshop, Universal Design for Learning, and Social and Emotional Learning   1

Chapter 2. Design Options for Welcoming Interests and Identities   33

Chapter 3. Design Options for Sustaining Effort and Persistence   61

Chapter 4. Design Options for Emotional Capacity   87

Chapter 5. Design Options for Perception   111

Chapter 6. Design Options for Language and Symbols   139

Chapter 7. Design Options for Building Knowledge   169

Chapter 8. Design Options for Interaction   201

Chapter 9. Design Options for Expression and Communication   227

Chapter 10. Design Options for Strategy Development   251

Chapter 11. You've Got This!   273

References 277

Index 287

 Visit the companion website at
https://companion.corwin.com/courses/MathWorkshopPlus
for downloadable resources.

Note From the Publisher: The authors have provided video and web content throughout the book that is available to you through QR (quick response) codes. To read a QR code, you must have a smartphone or tablet with a camera. We recommend that you download a QR code reader app that is made specifically for your phone or tablet brand. Videos may also be accessed at the URLs provided.

Preface

It is with great excitement, pride, and anticipation that we bring you this book. It has been a labor of love spanning multiple years, countless revisions, and significant input from respected colleagues and experts in the field. During this time, the Universal Design for Learning (UDL) Guidelines were revised, new research emerged, other books on UDL and math were published, and we gained more experience with many of the strategies presented in this book.

Before you begin, we want to take a moment to contextualize our work. This book deals with the intersection of huge ideas in the field of education, and we will be exploring them through a very specific lens. We acknowledge that we do not explore all facets of each area and want to be clear that our definitions are not all encompassing. This is probably most significant in relation to our exploration of equity.

Equity in education is an area that has been widely discussed, defined, and debated in recent years. Many authors have made significant contributions to the research base and educational landscape by writing about equity in significant depth. These authors have examined issues of race, culture, systemic marginalization, and the impact of these things on students within the context of classrooms. We applaud their work and have learned immeasurable lessons from their publications. If you are interested in taking a deep dive into those issues of equity, we highly recommend that you look to their body of work. To get you started, we have shared a few of our favorites, though there are many others. While we may touch upon some of the themes they examine, this book is by no means a comprehensive guide to equity, nor does it claim to be.

> **RECOMMENDED READING**
> - *Unearthing Joy: A Guide to Culturally and Historically Responsive Curriculum and Instruction*, Gholdy Muhammad
> - *Culturally Responsive Teaching and The Brain: Promoting Authentic Engagement and Rigor Among Culturally and Linguistically Diverse Students*, Zaretta Hammond
> - *Ruthless Equity: Disrupt the Status Quo and Ensure Learning for ALL Students*, Ken Williams

In our context, we will explore equity through the lens of accessibility—specifically as it relates to the teaching and learning of mathematics in grades K–8. Our hope is that the Math Workshop Plus model, which we introduce through this book, creates avenues for accessibility that may not typically exist in a classroom that favors whole-group instruction. Our intent is to elaborate on what is commonly known as the Math Workshop model by deliberately viewing it through the dual lenses of UDL and social and emotional learning (SEL). As we unpack the elements of UDL and SEL in relationship to math instruction and the workshop model, we will explore how students engage in mathematics within the classroom learning environment, with the goals of removing barriers that limit or prevent access to students and increasing the social and emotional growth they achieve through their mathematics learning.

We aim to offer as many strategies as possible to help you foster equitable access for all students. As you read, we invite you to brainstorm additional ideas for any area you feel needs more attention and encourage you to consult the experts in areas where we do not offer as much guidance. This book sits on a student-centered pedagogical model, and we hope that the ideas presented give all students opportunities to have their voices heard, find their mathematical power, and know that no matter how they show it, they are capable doers of mathematics.

We also want to add a note about the design of this book. Given that our work is situated in the tenets of Universal Design for Learning, we and Corwin have worked hard to make this book as accessible as possible. From the structure and layout to the fonts to the colors and use of illustrations and alt-text, we have followed accessibility best practices to ensure that this book can be read by all educators and student teachers in all formats, whether they have visual impairments like color-blindness, cognitive challenges, learning disabilities, or use assistive technologies.

Acknowledgments

Writing this book has been a true test of endurance, from an idea born during COVID-19 to the numerous drafts and revisions. I am eternally grateful to Erin Null for her calm guidance and support—this book wouldn't exist without her. To my partner, Dr. Nicki Newton, your friendship and collaboration mean everything. To my husband Jon, thank you for supporting my passion and for understanding the time and dedication it requires. To my daughters, Alaina and Julia, being your mom will always be my biggest achievement. Finally, to my wonderful family and friends, your patience and understanding when my work demands my attention are greatly appreciated. Thank you all sincerely for your continued support. I'm so grateful to all of you.

—Alison J. Mello

It takes a village to write a book. I thank the village. First, I thank God for the wisdom to write a book. I thank my family and friends who support me continually. I thank every teacher and student I have ever worked with because they inform my practice every day. I thank Alison, first for coming up with this idea and then inviting me on the journey. I thank Erin for taking the trip with us! Her patience is indescribable! I thank all of the behind-the-scenes staff at Corwin who actually make the book come to life, the copy editors, the art editors, the formatting folks, and all. It has been a pleasure and an honor to take this journey.

—Nicki Newton

Publisher's Acknowledgments

Corwin gratefully acknowledges the contributions of the following reviewers:

Kimberly Rimbey
Author, Corwin
Chief Learning Officer & CEO, KP Mathematics
Facilitator, Building Thinking Classrooms
Glendale, AZ

Skip Tyler
Mathematics Consultant
Collaborative Teaching and Learning Group (CTLG) Consulting
Chestnut Drive, VA

Ann Elise Record
Consultant, Ann Elise Record Consulting LLC
Concord, NH

Katherine Ariemma Marin
Assistant Professor, University of Louisville
Louisville, KY

Rosalba Serrano
Zenned Math
Highland Mills, NY

Lori Breyfogle
Elementary Math Specialist
Imperial, MO

Janet D. Nuzzie
District Intervention Specialist, K–12 Mathematics, Pasadena
  Independent School District
Pasadena, TX

Carly Morales
Instructional Coach, School Dist. 93
St. Charles, IL

About the Authors

Dr. Alison J. Mello has been in education for 30 years as a classroom teacher, math specialist, director of curriculum, and assistant superintendent. She is a coauthor of *Fluency Doesn't Just Happen With Addition and Subtraction, Fluency Doesn't Just Happen With Multiplication and Division, Working With the Beaded Number Line, A Teacher's Guide to Math Workshop,* and the forthcoming *Working With One-Inch Tiles* and *Building Better Math Instruction: A Toolkit for Leaders.* She has spent the last decade as a national speaker, math consultant, and graduate instructor of inservice and preservice teachers. Alison enjoys developing practical strategies to address issues that districts and teachers face in their schools and classrooms every day. She uses her understanding of leadership, curriculum, and best practices to inspire and assist leaders and teachers in elevating and transforming math instruction.

Dr. Nicki Newton is an education consultant who works with schools and districts around the country and Canada on K–8 math curriculum. She has taught elementary school, middle school, and graduate school. Dr Nicki has an EdM and an EdD from the Department of Curriculum and Teaching at Teachers College, Columbia University, specializing in Teacher Education and Curriculum Development. She is greatly interested in teaching and learning practices around the world and has researched education in Denmark, Guatemala, and India. She has written over 40 books and is excited to be part of the team of writers for the McGraw-Hill new series, *Reveal Math*. Her latest books include *Accelerating K–8 Math Instruction* (Teacher's College Press) and also *A Teacher's Guide to Math Workshop* (Newton, Mello, & Nuzzie through Heinemann). She is an avid pinner, blogger, and tweeter. Dr. Nicki will zoom into any book study group to chat; email her at drnicki7@gmail.com.

Chapter 1

The Big Ideas of Math Workshop, Universal Design for Learning, and Social and Emotional Learning

> **Question to Teachers: What is your biggest challenge?**
>
> **Teachers: TIME. There are not enough hours in the day to meet the needs of students!!**

If you are a teacher, you know some hard truths. Each year, you receive a new batch of students and often add or lose a few during the year. There always seems to be a new program, and there are countless initiatives being cast your way, which you may or may not understand. You have so many standards to teach, and there's never enough time in the day to get everything done. You work hard, but the results don't always reflect your efforts. When this happens, it's frustrating.

For many, teaching is a labor of love. It's a continuous cycle of learning, and every day and year is different. Teaching has never been simple, and in recent years, it's become more complex. Classrooms are filled with students who have unique needs, varied experiences, exposures to different types and levels of trauma, and a wide range of strengths and challenges. Diversity can take many forms:

- Neurodiversity
- Racial Diversity
- Socioeconomic Diversity

- Gender Diversity
- Cultural Diversity

Reframing differences as assets helps to reveal the strengths in all students. The first step is recognizing, embracing, and leveraging the variability of our learners. When we embrace learner variability and create environments conducive to learning for all students, we educate the *whole child* (Darling-Hammond & Cook-Harvey, 2018). Key findings in the research related to teaching the whole child indicate that:

- Development is malleable.
- Variability in human development is the norm, not the exception.
- Human relationships are the essential ingredient that catalyzes healthy development and learning.
- Adversity affects learning—and the way schools respond matters.
- Learning is social and emotional, as well as academic.
- Children actively construct knowledge based on their experiences, relationships, and social contexts.

To successfully teach the whole child, we must pay special attention to the learning environment we create. You can start by asking yourself the questions in Figure 1.1.

Figure 1.1 • Questions to Help Teachers Foster a Holistic Learning Environment

- How can I cultivate opportunities for students to grow as members of a community?
- How can I maximize my instructional minutes?
- How can I meet the needs of ALL learners?
- How can I create an inclusive learning environment?

Source: icons from left to right: by istock.com/Vector DSGNR; istock.com/Maria; istock.com/KRdesign; istock.com/bortonia

Students *need* more, parents *expect* more, schools and districts *demand* more, and there is only *one of you.* The need to make instruction more *efficient, effective,* and *easy* is real, and making it more *equitable* is one of the keys. These 4 Es will serve as our beacon as we reimagine how you and your students experience math.

What Is Math Workshop? Why Should I Use It?

> Since we started implementing Math Workshop, the biggest change that I've noticed is the students' love for math! I see fist pumps . . . excitement . . . joy. The kids love it! It's freeing to them; it's freeing to the teachers. Being able to choose what they want to work on as opposed to a certain workbook page has changed their worlds. They are playing games to practice. They are talking about math in ways that I never imagined. Sometimes I have kids just come and tell me that they love math, just to tell me, and that is so cool!
>
> We had our highest achievement on state testing ever, and I 100% believe that it is because of the Math Workshop model. Students are getting what they need, being challenged at the right levels, and feeling successful every day!
>
> —Meghan Zwolenski (math specialist, Massachusetts)

Math Workshop is a student-centered instructional approach designed to meet students where they are, maximize your minutes, and foster independence. It can be paired with any math curricular program and is effective whether your math block is 45 minutes or 90. It won't look the same in all grades and classrooms, but the structure within a lesson consistently includes four elements:

- Interactive Launch (5–10 minutes)
- Whole-Group Mini Lesson (10–20 minutes)
- Small-Group Instruction and Workstations (time varies)
- Whole-Group Debrief

Math workshop on its own is not new. You may already be quite familiar with it. We believe that Math Workshop holds the key to getting more done, in less time, with greater outcomes for all students. Math Workshop empowers you to maximize math minutes through small-group instruction and differentiated workstations. By beginning with accessible routines, you cultivate an environment that develops positive math dispositions and encourages risk-taking.

> **We believe that Math Workshop holds the key to getting more done, in less time, with greater outcomes for all students.**

Rather than "one-size fits all" instruction, Math Workshop offers options through menus, playlists, choice boards, or station rotations. Math Workshop actualizes Vygotsky's social learning theory and situates learning in the Zone of Proximal Development (Vygotsky, 1978) for each learner. Within Math Workshop, students can do all these things:

- Set goals
- Engage in purposeful practice
- Collaborate with peers
- Experience personalized instruction

Math Workshop provides opportunities for intervention, extension, acceleration, practice, and independence. It makes math interactive, differentiated, engaging, and fun!

What Is Math Workshop *Plus*?

As mentioned, Math Workshop is not new. Known by some as Guided Math, it gained mainstream popularity around 2009 with the publication of *Guided Math: A Framework for Mathematics Instruction* (Sammons, 2009). On its own, Math Workshop is an excellent way to differentiate instruction and help students develop a deeper understanding of math concepts, but it has always felt like it was missing a particular intentionality when it comes to equity.

Math Workshop Plus is therefore additive—it is designed with equity, access, and *excellence for all students* in mind. It is an elevated version of Math Workshop that infuses Universal Design for Learning (UDL) and social and emotional learning (SEL). As illustrated in Figure 1.2, Math Workshop Plus aims to instruct the whole child. It grows from the desire to humanize math and make it accessible for all. It considers the role of SEL competencies in learning. It recognizes that students learn differently and commits to teaching mathematics without ceilings, labels, or limits. It provides a learning environment that is safe and equitable.

Figure 1.2 • Moving From Math Workshop to Math Workshop Plus

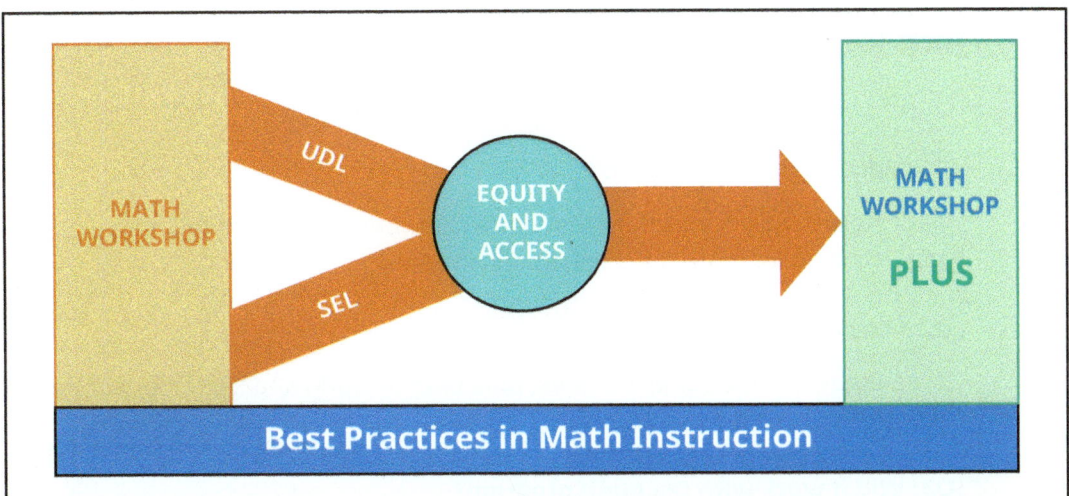

Adding this equity lens ensures that when Math Workshop is used, it gives all students greater access to grade-level content while *simultaneously building capacity in lagging skills.* In other words, *Math Workshop Plus* is an equitably designed space to build capacity in math *and* in language, social skills, self-regulation, time management, and personal responsibility.

Math Workshop enacts best practices and challenges ideas traditionally associated with math instruction. It stands in stark contrast to timed tests, answer-getting, and one-size-fits-all instruction. It shifts the focus from the teacher to the students and is active rather than passive. Unlike "drill and kill," thinking, talking, playing, questioning, and understanding are encouraged and valued.

When we add UDL to Math Workshop, we amplify the impact by cultivating a more inclusive environment and more equitable experiences. UDL removes barriers and creates opportunities to fortify skills. It leverages student strengths and recognizes proficiency in different forms. A universally designed Math Workshop also embeds opportunities to build and practice social and emotional competencies. Students negotiate partners, take turns, help, share space and materials, communicate, discuss, and the list goes on. It's Math Workshop PLUS a whole lot more!

Before adding the "plus," let's unpack the elements of Math Workshop, UDL, and SEL individually so that you can see them each on their own and then how they come together. Whether Math Workshop is new to you or you've been at it for years, we are honored that you are here and can't wait to help you design the most effective, accessible, and joyful math classroom you've ever led!

Math Workshop 101

Let's start with some FAQs about Math Workshop:

Q1: What is Math Workshop?

Math Workshop is a structure that breaks up your math block to create opportunities for more differentiated instruction.

Q2: Why do it?

It maximizes instructional minutes, puts data to work, and better meets the needs of your students.

Q3: Will it work with my math program?

Yes! (more on this later)

Q4: What do I need to get started?

The right mindset and some basics like cards, dice, manipulatives, and simple games.

Q5: Will this be a lot of work?

At first, but once you get organized, establish your systems, and get in a routine, it will be *less* work.

Q6: Is Math Workshop ability grouping?

NO. This is one of the biggest misconceptions about Math Workshop (more in Figure 1.2).

Q7: If Math Workshop can better meet the needs of all students and work with any math program, what stops some people from trying it?

We believe that misconceptions about Math Workshop may be the issue. Let's explore Table 1.1 to clarify what Math Workshop is and what it is not.

Table 1.1 • What Math Workshop Is and Is Not

Math Workshop Is	Math Workshop Is Not
Flexible	Ability grouping
Student-centered	Teacher-centered
Data-informed	Random
Discourse rich	Teacher talk
A structured approach	A program
Differentiated	One-size-fits-all

Key Components of Math Workshop

Math Workshop is not a program, a fad, or ability grouping in a pretty package. It's an instructional model designed to make instruction more efficient and effective. Math Workshop condenses whole-group instruction so you have time to meet with students in small groups, and there is more time to develop and practice essential skills. It chunks your math block into four clearly defined parts that support engagement, independence, and differentiation.

Opener

Math Workshop opens with a routine or energizer that gets students thinking, talking, and engaged. It lasts a few minutes, promotes reasoning and discussion, and allows all students to begin with a positive experience. It also offers a chance to listen to what students know and understand. As opposed to a typical "do now" where students silently push a pencil, students begin their math experience with enthusiasm, energy, and success!

Mini Lesson

The mini lesson is the whole-group instruction component of Math Workshop. Focusing on the learning target keeps it 10 to 20 minutes in length. While this takes practice, it helps to remember that the goal is not mastery for all students in that moment. Since Math Workshop includes small-group instruction and a variety of practice opportunities, students who need more time will receive it, and those who are ready to move on can do so.

> If keeping lessons mini is challenging, shift to focusing on the students who *do* understand rather than those who *do not*. These students often have their needs neglected—and sometimes exhibit challenging behaviors out of boredom—because they sit through lessons that they don't need. A good rule of thumb is to wrap it up when you see two-thirds of your class has the concept.

Workstations and Guided Math Groups

Following the lesson, students may complete a "must do," such as an exit ticket or practice page from your instructional materials, before transitioning to Guided Math groups and workstations. To run groups without interruption, students must know what to do, who to work with, how to check their work, and what to do if they get stuck. Through a mix of games, task cards, and technology, students can revisit prior learning regularly, which results in greater retention and transfer (Agarwal et al., 2021; Emeny et al., 2021). Stations can also offer practice with current content, preview future learning, and support vocabulary. Workstations give students what they need, when they need it, and often allow them to make choices and dictate their own pace.

Guided Math groups may be heterogeneous or homogeneous and should meet a specific learning target. The formation of these groups is informed by real-time data, making them fluid and flexible. While a group *may* focus on the current lesson, another group may revisit a prior skill, explore an extension, or fill in prerequisite skills. Workstations allow learning to continue even when students are not working directly with you, so you need not see every student in a small group every day. Strive for balance so you meet with all students throughout the week. This will help avoid the tendency to see just certain students. Using data is critical to ensure that students don't remain in the same groups for long. This is not the goal and should be avoided.

Debrief

Math Workshop closes with a debrief. This is sometimes called the "share" because students share what they learned. It takes 2 to 5 minutes and allows you to hear, directly from students, what they learned. It's also a chance to see if there are any lingering misconceptions and to come back together as a community of learners. During the debrief, we love seeing students share affirmations and encourage each other.

If it's helpful to have a quick reference guide for each component of Math Workshop, take a quick photo of Table 1.2. It includes descriptions, examples, and reminders.

Table 1.2 • Key Components of Math Workshop

Component	Examples/Definitions	Minutes
Routine or Energizer (Interactive Launch)	Examples: • Number Talks, I Have Who Has, Guess My Rule, Target Number, True or False, Which One Doesn't Belong	5–10
Mini Lesson	• Whole-group lesson • Focused on specific learning target • Involves some form of explicit instruction • Student-centered and hands-on when possible	10–20
Guided Math Groups	• 3–5 students • Groups change often ○ Learning targets informed by current data • Intervention, extension, reinforcement, acceleration	Time varies (5–15 min per group)

Workstations	• Students choose based on their goals
	• Purposeful practice
	  ○ Rigorous
	  ○ Essential standards
	• Differentiated
	• Games, projects, adaptive technology
	• Self-checking
	• Artifacts (for accountability)
		Time varies

Debrief (The "Share")	• Return to whole group
	• Share learning of the day
	• Tips to keep it brief
	  ○ Have students turn and talk first
	  ○ Select 4–5 students to share with the whole group (equity sticks are great for this)
	  ○ When short on time, have students write on a sticky note or in their journal
		2–5

Although the components are consistent, Math Workshop looks and feels different in every classroom. Find what suits you, such as station rotation versus menu, which will be explained more later. Determine what noise level works and what student accountability will look like. Through trial and error, find the right organizational system, how much choice is comfortable, and how you'll manage paper and supplies. Decide how to take notes during Guided Math groups and how frequently to change workstations (we recommend weekly or biweekly). Experiment with routines, practice keeping lessons mini, and try the debrief. Once you feel good, you're ready for the *plus*.

Math Workshop Plus takes you to the next level. It aims to close opportunity gaps and open doors. It requires tenacity, planning, and your belief that every student can and will succeed when the right supports are embedded in the learning environment. The magic of Math Workshop Plus comes from creating conditions for all students to effectively communicate their thinking, strategies, and solutions. It comes from ensuring resources and tools are readily accessible. It comes from anticipating student needs and planning for them. The magic is UDL, and it will change your workshop in all the best ways.

What Is Universal Design for Learning?

> Universal Design for Learning (UDL) is a teaching approach that works to accommodate the needs and abilities of all learners and eliminates unnecessary hurdles in the learning process. This means developing a flexible learning environment in which information is presented in multiple ways, students engage in learning in a variety of ways, and students are provided options when demonstrating their learning.
>
> *Cornell University Center for Teaching and Innovation*

Universal Design is a concept developed by Ron Mace of North Carolina State University. His vision grew from the architecture world as a method to design products and environments that met the conditions of the Americans with Disabilities Act (ADA) while also serving everyone without making adaptations. Examples include curb cuts and large restroom stalls. Both were designed to accommodate wheelchairs but also serve strollers, shopping carts, and more. As the term implies, the design is universal, so the function can help everyone.

Universal Design for Learning (UDL) is a framework inspired by Mace and based on the work of Vygotsky (1962). In education, accessibility refers to many things, but let's take a simple example. Many classrooms install amplification systems for those with impaired hearing. These systems also benefit students who struggle with attention, have auditory processing issues, or sit in the back of the classroom. They even help teachers preserve their voices! Everyone benefits without the need for additional adaptations. The result is increased accessibility for more students without additional work for the teacher. Less work with better outcomes? Yes, please!

The UDL framework is divided into three principles, which outline how the three networks of the brain work together when learning. The domains examine these three characteristics:

1. Recognition—the *what* of learning
2. Skills and strategies—the *how* of learning, and
3. Caring and prioritizing—the *why* of learning.

In simple terms, effective learning involves knowledge, skills, and enthusiasm, and putting them all together is the recipe for success.

In July 2024, the Center for Applied Special Technology (CAST) debuted the UDL Guidelines 3.0. As seen in Table 1.3, the goal is learner agency. The graphic is structured around Engagement, Representation, and Action & Expression, each with specific guidelines and considerations to elevate access, support, and executive function. This book will touch upon most of them. Explore Table 1.3 and consider how the guidelines and considerations align with the vision of Math Workshop Plus. You can see CAST's version of this table at https://udlguidelines.cast.org/.

Many ideas in the framework have the potential to improve outcomes in math, yet in 2021, Lambert et al. found that the framework itself was not enough to transform teaching and learning. They noted that "the potential of UDL rests in its power to redesign classrooms, curriculum, and systems to work better for students with disabilities" (p. 57) and all other students as well. After exploring the connection between design thinking and UDL in relation to mathematics education, they concluded that UDL is general. Thus, applying it in content areas requires intentional action. To support this, Lambert (2021) developed the UDL Math Design Elements framework to connect the principles of UDL with mathematics education research. As seen in Table 1.4, the hope is to shift the perception of UDL as a "thing," or a noun, to an "action," or a verb. Many ideas in the graphic intersect with Math Workshop and SEL, and there is a strong theme of equity. This graphic may help to synthesize ideas within the chapters, so you may want to flag this page.

Table 1.3 • Universal Design for Learning Guidelines 3.0

Design Multiple Means of **Engagement** 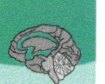	Design Multiple Means of **Representation** 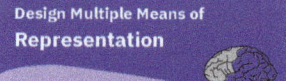	Design Multiple Means of **Action & Expression** 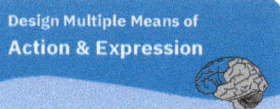
Access		
Design Options for **Welcoming Interests & Identities** • Optimize choice and autonomy • Optimize relevance, value, and authenticity • Nurture joy and play • Address biases, threats, and distractions	Design Options for **Perception** • Support opportunities to customize the display of information • Support multiple ways to perceive information • Represent a diversity of perspectives and identities in authentic ways	Design Options for **Interaction** • Vary and honor the methods for response, navigation, and movement • Optimize access to accessible materials and assistive and accessible technologies and tools
Support		
Design Options for **Sustaining Effort & Persistence** • Clarify the meaning and purpose of goals • Optimize challenge and support • Foster collaboration, interdependence, and collective learning • Foster belonging and community • Offer action-oriented feedback	Design Options for **Language & Symbols** • Clarify vocabulary, symbolic, and language structures • Support decoding of text, mathematical notation, and symbols • Cultivate understanding and respect across languages and dialects • Address biases in the use of language and symbols • Illustrate through multiple media	Design Options for **Expression & Communication** • Use multiple media for communication • Use multiple tools for construction, composition, and creativity • Build fluencies with graduated support for practice and performance • Address biases related to modes of expression and communication
Executive Function		
Design Options for **Emotional Capacity** • Recognize expectations, beliefs, and motivations • Develop awareness of self and others • Promote individual and collective reflection • Cultivate empathy and restorative practices	Design Options for **Building Knowledge** • Connect prior knowledge to new learning • Highlight and explore patterns, critical features, big ideas, and relationships • Cultivate multiple ways of knowing and making meaning • Maximize transfer and generalization	Design Options for **Strategy Development** • Set meaningful goals • Anticipate and plan for challenges • Organize information and resources • Enhance capacity for monitoring progress • Challenge exclusionary practices

Source: Adapted from CAST, 2024. Universal design for learning guidelines version 3.0. Retrieved from https://udlguidelines.cast.org. Used with permission.

Table 1.4 • UDL Math Design Elements

Strategic sense makers: Are students constructing identities as strategic sense makers in math?		
Engagement	Meaningful mathematics	Is the math in your class meaningful to students?
		Do students regularly engage in sense making?
	Supportive classroom environments	Do your students feel safe enough to take mathematical risks?
		Are they building relationships in and through math?
Representation	Multimodal	Is math content accessible? Multimodal?
		Can students choose how they solve problems?
	Focus on core ideas	Does the design of your class guide students to understand and remember core mathematical ideas?
Strategic action	Equitable feedback	Does feedback help students grow as mathematicians?
		Is assessment equitable for all learners?
	Understanding self as a math learner	What do your students learn about themselves as math learners?
		How do you support that development?

Source: Adapted from Lambert, 2021. Used with permission.

Why a Universally Designed Math Workshop?

The goal of UDL is "learner agency that is purposeful and reflective, resourceful and authentic, strategic, and action-oriented" (CAST, 2024). This goal is aligned with Math Workshop, where we cultivate conditions to support student collaboration, independence, and engagement. Reflect on the learning environment in your classroom. How motivated are students? What do they know about how they learn? Do they believe math is about sense-making, or answers? Can they utilize resources such as anchor charts and math tools? Do they? Do they self-start or wait to be guided through steps? Are there opportunities for choice? Do they know where they are on the learning continuum? Are they attuned to what they know? Do they know what they need to learn? Do you?

In Math Workshop, students are clear on their purpose, and teachers are comfortable giving students more control in the classroom. In this environment, choice is normalized, and agency is fostered. Motivation improves when students feel in control of their learning (Code, 2020) and can decide what to do, where to do it, and for how long. Choice is a cornerstone of UDL, and that extends to having options to demonstrate proficiency. Typically, math proficiency is measured by written work, especially paper-and-pencil or computer-based assessments. Within Math Workshop Plus, students may demonstrate proficiency in a variety of ways.

Agency is fostered as students access and use the resources and tools they prefer, as they persevere and work with others to overcome obstacles and also as they set goals and monitor their progress. Students know where they are on the learning continuum, where they are going, and why. We enter these classrooms and ask students not what they are *doing* but what they are *learning*. Try it and you may be surprised by the responses!

Adding UDL to Math Workshop means anticipating learner variability and planning to remove barriers (Murawski & Novak, 2019). It means being *proactive* versus *reactive.* If this sounds like more work, chances are you already include some UDL elements, so taking your practice to the next level could be as straightforward as shifting how you prep and plan. Teachers who plan with a UDL lens ask themselves questions that help them provision for differences. This increases accessibility for all students and decreases the need for individual modifications, which results in *less* work.

Try It!

Proactively Designing Options for Engagement, Representation, and Action & Expression

Try this short exercise using the bullet points in Table 1.3. Think "How can I proactively design options for (fill in the blank)?":

Engagement

- ✓ Welcoming interests and identities
- ✓ Sustaining effort and persistence
- ✓ Emotional capacity

Representation

- ✓ Perception
- ✓ Language and symbols
- ✓ Building knowledge

Action & Expression

- ✓ Interaction
- ✓ Expression and communication
- ✓ Strategy development

If we expect students to own their learning during Math Workshop Plus, it is helpful to pause and reflect on where they often get stuck. When we do this, we can proactively adjust. This prevents them from getting stuck in the future, gives them confidence when confronted with other barriers, and builds stamina, perseverance, and independence.

Whether you are already engaged in this work, just getting started, or thinking about it, this book is filled with practical, research-informed, and easy-to-implement ideas to help maximize every opportunity to support all learners. As you think about your students and your classroom, consider the following questions, and add others you may have. Jot them down and be on the lookout for answers as you read.

> **Try It!**
>
> Reflecting on My Classroom
>
> Can all students access the language?
>
>
>
> How can I make the concept more visual?
>
>
>
> What tools can students use to support their thinking?
>
>
>
> Is there more than one way to demonstrate proficiency?

What questions should you consider when planning a universally designed lesson, workstation, or Guided Math group? CAST has developed the following guide to activate your thinking about learner variability and spark ideas to proactively plan for differences. After reviewing Exhibit 1.1, we encourage you to jot down your thoughts in the space provided and share them on your social media platform of choice using #mathworkshopPLUS.

Exhibit 1.1 • Key Questions to Consider When Planning Lessons

Think about how learners will engage with the lesson.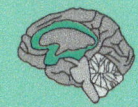

Does the lesson provide options that can help all learners:
- regulate their own learning?
- sustain effort and motivation?
- engage and interest all learners?

Think about how information is presented to learners.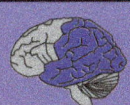

Does the information provide options that help all learners:
- reach higher levels of comprehension and understanding?
- understand the symbols and expressions?
- perceive what needs to be learned?

Think about how learners are expected to act strategically and express themselves.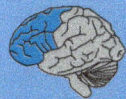

Does the activity provide options that help all learners:
- act strategically?
- express themselves fluently?
- physically respond?

Source: Adapted from CAST, 2024. Universal design for learning guidelines version 3.0. Retrieved from https://udlguidelines.cast.org. Used with permission.

So far, we have explored the connections between Math Workshop and UDL, but you may have noticed a connection to social and emotional learning (SEL). We certainly did!

What Is Social and Emotional Learning (SEL)?

> Social and emotional learning (SEL) is an integral part of education and human development. SEL is the process through which all young people and adults acquire and apply the knowledge, skills, and attitudes to develop healthy identities, manage emotions and achieve personal and collective goals, feel and show empathy for others, establish and maintain supportive relationships, and make responsible and caring decisions.
>
> SEL advances educational equity and excellence through authentic school–family–community partnerships to establish learning environments and experiences that feature trusting and collaborative relationships, rigorous and meaningful curriculum and instruction, and ongoing evaluation. SEL can help address various form of inequity and empower young people and adults to co-create thriving schools and contribute to safe, healthy, and just communities.
>
> *CASEL, 2020*

You are likely familiar with the SEL competencies developed in 1997 by the Collaborative for Academic Social and Emotional Learning, better known as CASEL. The presence or absence of these competencies can color the culture and climate of your classroom, and lagging social and emotional skills can inhibit academic potential (Immordino-Yang, 2016). Let's examine the SEL competencies through the lens of the Math Workshop Plus and identify opportunities to strengthen these skills.

Math Workshop Plus aspires to create a more accessible and equitable learning environment for all students. In 2020, CASEL refined their definition of social and emotional learning to highlight the role of SEL in fostering equity in schools and classrooms. The revised definition—shared at the beginning of this section—also emphasizes "student agency," which is the goal of UDL and a key element of Math Workshop.

Did you make any connections? If access and equity are the "plus" we are adding, our classrooms and learning experiences must reflect that. As we make decisions, prepare lessons, and build our toolboxes, we should be thinking of our students as individuals. By design, Math Workshop does this. It moves us away from one-size-fits-all and empowers students to own their learning. Since this may be a new experience for students, we must provide guidance and support. If we remind ourselves that they are still in the process of developing as humans *and* as learners, we can leverage Math Workshop Plus to build and strengthen SEL competencies.

> **Try It!**
>
> Connecting Math Workshop
> Plus and SEL
>
> As you read the definition of SEL, what connections can you make to Math Workshop Plus? It may be helpful to revisit Figure 1.2 as you examine how UDL and SEL come together to foster the access and equity that put the "plus" in Math Workshop Plus.

We believe that *all* learning is social and emotional. Much of our growth as humans comes from lessons learned through our exchanges with peers, friends, and the people we interact with daily. Returning to the idea of teaching the whole child, Darling-Hammond and Cook-Harvey note (2018) that "human relationships are the essential ingredient that catalyzes healthy development and learning." The following quote from Dr. James Comer, shared by Rita Pierson in her viral TED Talk (2013), captures this perfectly:

"No significant learning happens without a significant relationship."

Math Workshop is fertile ground for relationship building. It positions you to work in closer proximity to students in Guided Math groups, which allows you to observe student thinking in real time, discuss conceptions as they develop, and address misconceptions before they grow. Through skilled questioning, students reflect, ponder, and revise their thinking. Teachers who implement Math Workshop often report feeling more effective, efficient, and empowered to support students as individuals. Students report feeling seen, heard, and valued as their teachers highlight what is *right* about their thinking while pushing them to persevere through challenges. This is a stark contrast to handing in a paper and getting it back days or weeks later, with red ink indicating wrong answers.

As you run Guided Math groups, the other students engage in workstations. Many teachers are skeptical about this and reluctant to offer choice. There is a genuine concern, and often for good reason, that autonomy within workstations will devolve into chaos. While possible, we know from years of experience and success that, when expectations are clear, routines are explicitly taught and rehearsed, and accountability is high, workstations offer effective opportunities to practice math and much, much more. This component of Math Workshop develops, strengthens, and applies SEL skills.

What are these skills, and how do they fit into Math Workshop Plus? As outlined in Figure 1.3, the SEL competencies include self-awareness, self-management, responsible decision-making, relationship skills, and social awareness.

Figure 1.3 • CASEL Social and Emotional Learning Framework

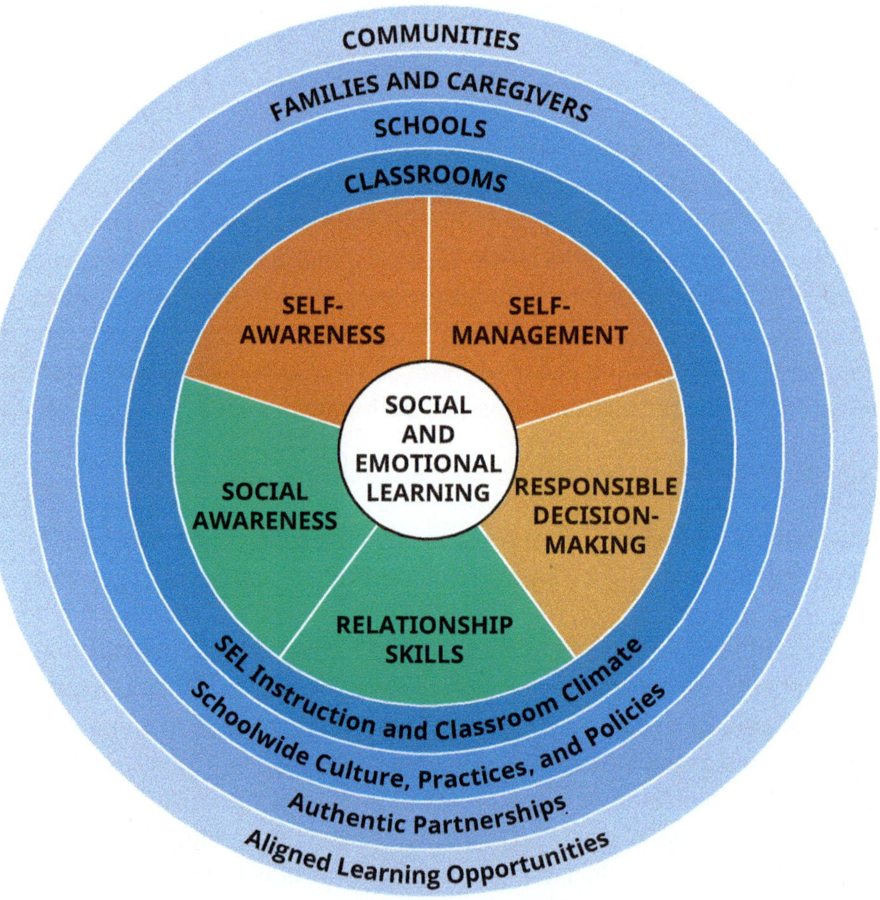

Source: CASEL, 2020. For more explanation, the framework is available for free at https://casel.org/casel-sel-framework-11-2020/.

Let's zoom in on each of the five competencies, understand what they mean, and think about how they can be developed, practiced, and strengthened through Math Workshop Plus.

As defined by CASEL in 2020, the five SEL competencies are as follows:

Self-Awareness

The abilities of students to understand their thoughts, emotions, and values, and how these things influence their decisions and behavior in different situations. Self-awareness encompasses the ability of students to recognize their strengths and limitations and possess a solid sense of confidence and purpose.

Self-Management

The abilities of students to effectively manage their emotions, thoughts, and behaviors in a variety of situations to achieve their goals. Self-management encompasses having the ability to manage stress, delay gratification, and have motivation and agency to reach individual and group goals.

Social Awareness

The abilities of students to show empathy and understand the perspectives of others, including people who are different from them. Social awareness includes feeling compassion for others, understanding social norms and behaviors appropriate for different settings, and accepting the resources and supports available from the school, community, and family members.

Relationship Skills

The abilities of students to form and maintain supportive, healthy relationships and effectively work with diverse people in different settings. Relationship skills encompass clear communication, active listening, cooperation, collaborative problem-solving, and constructive conflict resolution. This includes navigating situations with different social and cultural expectations, offering leadership, helping others, and pursuing help for themselves when needed.

Responsible Decision-Making

The abilities of students to make constructive and caring choices related to personal behavior and social interactions in a variety of diverse situations. Responsible decision-making encompasses considering safety and ethical concerns and assessing the benefits and consequence of actions in relation to social, personal, and collective well-being.

We've heard from countless teachers that too often the "L" in SEL gets lost in the conversation and that SEL is often treated as a separate thing. Since research notes the role of these five key social and emotional competencies in learning (Dusenbury et al., 2020), our goal is to capitalize on opportunities *within* the context of learning rather than *outside* of it. Let's explore how the competencies promote or impede learning and highlight where they are utilized in Math Workshop Plus.

Opportunities to exercise SEL skills are abundant in Math Workshop. Workstations offer partner, group, and independent learning options. They allow students to practice how to speak with each other, how to respect the thinking of others, and how to build their decision-making muscle. Students must make choices to progress along the learning trajectory. They must regulate their emotions when they lose a game, wait their turn, or can't work with their desired partner. Students are *trusted* to oversee their learning and experience authentic outcomes of their actions and decisions.

As you familiarize yourself with Math Workshop Plus, consider what role this model can play in developing SEL competencies. Throughout each chapter, we will note connections to SEL and offer strategies to cultivate and maximize opportunities within this enhanced universally designed version of Math Workshop.

A Glimpse of SEL in Action at the Launch of Math Workshop Plus

Here is a glimpse into a Math Workshop Plus launch that illustrates how the classroom community looks, feels, and sounds. The teacher begins with this Which One Doesn't Belong routine (Figure 1.4).

Figure 1.4 • Which One Doesn't Belong

Source: Andrew Gael, WODB talkingmathwithkids.com.

Mrs. Kye:	Decide which coins don't belong. Take a moment to notice and wonder and really think about it. Be prepared to explain WHY you made your selection.
	(allows for think time)
Cole:	I think the dimes don't belong.
Mrs. Kye:	Who agrees with Cole? (many students signaling that they agree) OK, Cole, why don't they belong?
Cole:	They don't belong because they total 50 cents, and the others total 25 cents.
Mrs. Kye:	Hmm . . . is that a valid argument? Do you agree? (students signaling that they agree) How can we check?
Maria:	Well, we can count the collections: 5 nickels are 25 cents; 25 pennies are 25 cents; a quarter is 25 cents; 5 dimes are 50 cents—he's right!
Mrs. Kye:	Do we agree? (yes, yes, yes) Did anyone make a different choice? (many hands go up) Carla, what did you pick?
Carla:	Well, I thought that the quarter did not belong.
Mrs. Kye:	Did anyone else pick that one? (students smiling and agreeing) OK, Carla, why did you select the quarter?
Carla:	Well, what I noticed was that it was the only one that had just one coin.
Mrs. Kye:	Hmmm . . . is that a true statement? Do you agree? (students signaling that they agree) Can both Cole and Carla be correct? Why?
Juan:	Yes, because they gave a good reason, a true reason why they didn't belong. I didn't pick those, but I see what they are saying, and I agree with them!
Mrs. Kye:	Oh . . . that makes sense. Wait, you said you didn't pick the dimes or the quarter, so which one did you think did not belong?
Juan:	I chose the pennies.
Mrs. Kye:	Hmmm . . . OK, everyone, turn and talk with your group or think to yourself why Juan may have chosen the pennies. See what reasons you can come up with.
Mirko:	Can we share?

Mrs. Kye:	Well, first let's see what Juan's reason was. Juan?
Juan:	Well, I noticed that they were a different color. The others are silver.
Mirko:	Yes!! That's what we thought too. At first, we noticed that they were brown, and all of the others were silver.
Mrs. Kye:	So . . . is that a valid argument? (class saying yes excitedly; Juan looks happy)
Mirko:	(excitedly) But we came up with two more reasons. Can we share them?
Mrs. Kye:	I wonder if other groups had the same ones? Did other groups come up with more than one reason for the pennies not to belong? (students very excitedly saying yes, yes!)
Mrs. Kye:	Should we let Mirko share and then see if there are any other ideas? (class says yes)
Mirko:	We noticed the pennies were the only ones in an array. Like, it's the only one with rows and columns (some saying that's what we said!)
Mrs. Kye:	Do you agree? Is that a valid reason? Do you see the rows and the columns? Where? (student shows) Is it an array? (class shows and agrees) Mirko, is your group all done? (yes) OK, do any other groups have a reason for the pennies?
Shelby:	Yes! We noticed they were arranged in equal groups—5 groups of 5.
Mrs. Kye:	Interesting . . . is that a true statement everyone? (class agreeing as pointing and chatter spreads)
Mrs. Kye:	OK, anything else with the pennies?
Nick:	Well, this isn't really mathy or anything, but we noticed the face was pointing right and the other faces are to the left.
Mrs. Kye:	Hmmm . . . you are really looking closely! Class, is that a true statement? (YES! Others noticing) Wow! Do you think there may be even more things we haven't noticed? (yes!!)
Yeshi:	(interrupting) Wait! We didn't talk about the nickels, and that's what I chose!!

Mrs. Kye: Hmmm . . . OK, class, without saying it out loud, decide why you think Yeshi picked the nickels. Take 10 seconds. (wait) OK, Yeshi, why did you pick the nickels?

Yeshi: I picked the nickels because they were the only ones on the tails side!! (class erupts . . . yesssss!!!)

Can you feel the excitement, engagement, and energy? Can you see how each idea was valued, honored, and shared? That is how a Math Workshop always begins. Students are thinking, wondering, sharing, and getting their brains activated! In this example, students didn't need knowledge of coins to participate. They could notice color, arrangement, size, or mathematical things like values, quantities, and equal groups. The routine is visual, auditory, and oral, and there isn't one "correct" answer.

Did you notice how students learned from each other? The academic vocabulary didn't come from the teacher. Instead, the teacher facilitated discussion by asking questions to move thinking forward. Rather than focusing on the answer, the teacher probed for student thinking. Students supported each other, and the classroom as a community of learners was strengthened. SEL competencies were developed, modeled, fortified, and practiced.

The example illustrates how students can authentically develop, practice, and strengthen their SEL skills all while engaging in math, particularly the math practices. When looking at the Standards for Math Practice through the lens of SEL, examples of opportunities to strengthen these competencies may jump out at you. As you encounter examples, consider how to incorporate them, maximize them, and how your students could benefit from them. Consider how a commitment to agency, UDL, and independence can influence math disposition and self-efficacy. Also consider the role of these things in creating an equitable learning environment. According to CASEL,

> SEL advances educational equity and excellence through authentic school–family–community partnerships to establish learning environments and experiences that feature trusting and collaborative relationships, rigorous and meaningful curriculum and instruction, and ongoing evaluation. SEL can help address various forms of inequity and empower young people and adults to co-create thriving schools and contribute to safe, healthy, and just communities.

SEL as a lever for equity may not be as obvious as UDL, but when you think about what students with lagging SEL skills often miss out on, it makes sense. Students with weak SEL skills often struggle with relationships with peers and adults, experience poorer academic achievement, and suffer consistent and sustained behavior problems (McDonald et al., 2018). Their behaviors may result in missed instruction, a negative disposition toward math, and/or a lack of motivation to persevere when faced with rigorous problems. As previously noted, we view equity as a call to action. We see Math Workshop Plus as a space designed for equitable access but know that to fulfill that promise we must constantly reflect on our decisions, systems, and expectations to ensure that all students have the tools to succeed within the learning environment.

How Does Math Workshop Plus Support Equity?

How do you define equity? How about your colleagues and administrators? Does everyone define equity the same way? What practices support equity? What practices impede it? How do we define equitable math instruction? What does equity look like in Math Workshop?

Equity is a word that means different things to different people, yet it is frequently spoken about in education as if everyone defines it the same way. For this reason, we want to examine equity through the lens of math instruction and develop a common definition to use for the purpose of this book. The National Council of Teachers of Mathematics (NCTM) notes the following practices that support equity:

> These practices include, but are not limited to, holding high expectations, ensuring access to high-quality mathematics curriculum and instruction, allowing adequate time for students to learn, placing appropriate emphasis on differentiated processes that broaden students' productive engagement with mathematics, and making strategic use of human and material resources. When access and equity have been successfully addressed, student outcomes—including achievement on a range of mathematics assessments, disposition toward mathematics, and persistence in the mathematics pipeline—transcend, and cannot be predicted by students' racial, ethnic, linguistic, gender, and socioeconomic backgrounds.

Consider the situations in the following list that describe "invisible" barriers to equity. How do they reflect the practices described by the NCTM? How might they result in opportunity gaps for some students? Do any of these situations exist in your classroom, school, or district?

1. Some teachers and classrooms within the same school/district have more manipulatives than others.
2. Some students receive pull-out math support while the rest of the class is having Math Workshop (or math instruction).
3. Some students meet with both the classroom teacher and the special educator or interventionist, leaving little time for games or other workstations.
4. Some students do not speak English, so they are given practice pages of computation problems, or they leave during math time.
5. Some teachers use different instructional resources than their colleagues, which may include different vocabulary, different models, and/or different assessments.
6. Some students have the opportunity to try "challenge problems," but others do not.
7. Students may only use the manipulatives that the teacher selects when the teacher distributes them for a specific lesson.
8. Some grade levels, schools, or classrooms have more instructional minutes designated for math than others.

Throughout this book, we propose elevating Math Workshop through UDL to minimize inequities. You may notice an inherent relationship between UDL and equity, but in recent years, there have been more intentional efforts to apply UDL to ensure access for diverse learners. As the needs of students, including SEL needs, continue to vary and expand, it is imperative that we intentionally design and maintain learning environments that create conditions for success.

> Equitable learning environments "respond to the diversity of a school or classroom community, intentionally create rich opportunities for student action and reflection, attend to young people's psychological experience of learning, and develop their feelings of competence, connectedness, and purpose." SEL implementation both contributes to and depends upon an equitable learning environment where all students and adults feel respected, supported, and engaged.

Is it becoming clear that Math Workshop Plus strives to create such environments? Creating learning spaces where everybody gets what they need is an ideal actualized through this model. In a high-functioning Math

Workshop, there are no labels or levels, no high groups or low groups, and no judgement or assumptions. It is a safe space for students to share ideas, think, persevere, collaborate, fail forward, and explore. It is a structure that values and embraces differences, champions individuality, and recognizes that competence and proficiency in math are not defined by speed but by evidence of understanding that may be demonstrated with objects, pictures, explanations, or in writing.

We believe that Math Workshop Plus teaches students to be respectful of the ideas shared by peers, values the thinking of all members of the classroom community, and celebrates connectedness by affirming when different people share the same ideas. The environment highlights that there is more than one correct way to approach and solve problems and encourages students to use strategies that work best for them. In this way, students become more metacognitive. When students understand themselves as learners, they can more effectively advocate for themselves and own their learning. As we explore ideas such as goal setting, progress monitoring, and learning trajectories, we hope you see that Math Workshop Plus is not about comparisons to other students or grades on report cards. Math Workshop Plus is about advancing the learning of each student by supporting all students on their individual journeys.

Math Workshop Plus is a powerful structure for all that:

- Simultaneously empowers both students and teachers
- Intentionally removes barriers to learning
- Ensures that all students get what they need
- Offers options for all types of learners
- Embraces variability as an asset to the learning community
- Focuses on the whole child
- Creates an environment of respect for all

Where Will We Go From Here?

What made you pick up this book? Are you reading it because you're a Math Workshop believer looking to take your practice to the next level? Are you reading it because you are curious about Math Workshop and want to know how it works? Maybe it was the SEL connection that piqued your interest, or perhaps you want to know more about UDL. Regardless of why you are here, this book provides all of that and more. Our goal is to help you maximize instructional minutes, meet the needs of all students, and sharpen your focus on accessibility, equity, and the role of social and emotional competencies in learning.

You may be wondering how so much information will come together in a way that makes sense and how prepared you will be to make this a reality in your classroom. Understanding how the book is constructed and how to use the tools within each chapter will help make the ideas actionable.

The design of this book takes you on a journey to build an understanding of the UDL Guidelines through the lens of Math Workshop. With each chapter, you will gain more insight into the specifics of each dimension of the UDL framework and discover concrete examples to enact them in your classroom. If that sounds daunting, you'll find similar examples and strategies in different chapters due to some overlap within the Guidelines. We hope that the repetition helps solidify ideas and feeds your action plan.

As a reminder, the UDL Guidelines are organized into three areas: Engagement, Representation, and Action & Expression, each of which have a set of three unique guidelines. A chapter will be dedicated to each guideline. For example, under Engagement, the first guideline is on Welcoming Interests & Identities, which is comprised of these considerations:

- Optimize choice and autonomy
- Optimize relevance, value, and authenticity
- Nurture joy and play
- Address biases, threats, and distractions

As illustrated in Exhibit 1.2, each chapter will zoom in on the considerations within each guideline and illustrate them through concrete examples, resources, and strategies to make them a reality in your math workshop. The figure is intended to support your understanding of how each chapter uses that structure and how it relates to the UDL framework, SEL, and equity.

Exhibit 1.2 • How This Book Is Organized

Math Workshop Plus is organized around the UDL framework.

The UDL Guidelines are organized into three principles:

Provide Multiple Means of		
Engagement	Representation	Action & Expression

Each of these principles is broken down into specific guidelines. For example, within the principle of Engagement, there are three guidelines:

Engagement
Welcoming Interests & Identities
Sustaining Effort & Persistence
Emotional Capacity

Each of these guidelines is broken into considerations. For example, the Sustaining Effort & Persistence guideline includes five considerations.

Engagement
Welcoming Interests & Identities
Sustaining Effort & Persistence
Clarify the meaning and purpose of goals
Optimize challenge and support
Foster collaboration, interdependence, and collective learning
Foster belonging and community
Offer action-oriented feedback
Emotional Capacity

Each chapter in Math Workshop Plus is dedicated to a specific guideline. Within the chapter, we will zoom in on the considerations and illustrate them using concrete examples, resources, and strategies to bring them to life in your Math Workshop. Our hope is that this intentionality creates a road map for you that frames the UDL Guidelines like small shifts, slight adjustments, or easy upgrades you can make to what you already do that will result in greater access and better outcomes for all students.

In addition to the UDL principles, each chapter will include callouts along the way—highlighted by icons—to highlight reflection moments to help you consider more about them:

- **Connections to SEL** and what it looks, sounds, and feels like in Math Workshop Plus.
-  **Equity Checks** to determine what creates equity in your classroom.
-  **Action Plans** to prompt you to take immediate action in your next lesson.
- 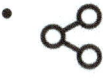 **Try It!** opportunities to engage in self-assessment related to the chapter.
- **Connection** moments, encouraging you to share what you are thinking, planning, or doing with the math Workshop Community, using #mathworkshopPLUS on your social media platform of choice.

Math Workshop Plus offers a dynamic approach to math instruction that aspires to build proficiency in math, while developing the whole child. By designing for multiple means of engagement, representation, and action and expression, and by enacting the UDL Guidelines to anticipate and remove barriers, students are empowered, confident, and play an active role in their growth. This alternative to a one-size-fits-all approach to instruction fosters inclusive, accessible learning environments for all learners. As students make decisions, interact with peers, build awareness of their thinking, and monitor themselves, they organically and authentically strengthen their SEL competencies. As you continually reflect on your actions, decisions, and systems, you become more attuned to the experiences of your students and how your classroom supports or inhibits equity and access.

Math Workshop Plus offers a space to maximize your impact and for students to be more independent, solve their own problems, and work with others. This model creates pathways for equity, access, and success for *all* learners within a safe environment where students flourish in their development as mathematicians and people. If you never have enough time, are challenged to meet a wide range of needs, and want to give every student what they need without compromising your personal sanity, this model is a dream come true!

Reflection Questions

1. How satisfied are you with your current math instruction?
2. How comfortable are you giving students more ownership over their learning?
3. What are your biggest takeaways from this chapter about the roles of SEL, UDL, and equity in math instruction?

 ## Action Plan

Assess your current math block. How do you use your minutes?

 ## Try It! Barriers and Opportunities

Make a list of your top five frustrations and identify which ones Math Workshop Plus could mitigate or eliminate.

Try a routine, such as Which One Doesn't Belong, and compare the experience with the example in the chapter.

 ## Connections

Start a journal to track your progress as you move through the book. On page one, write your reason for wanting to implement this model and what you are thinking right now. Commit to share your journey with a colleague or connect with someone on the same journey via social media using #mathworkshopPLUS.

Chapter 2

Design Options for Welcoming Interests and Identities

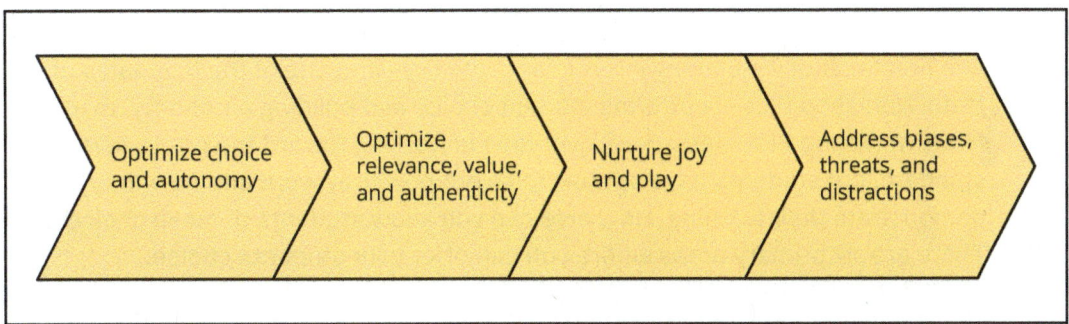

All students deserve to be engaged and interested in what they are learning, and for many, math is neither engaging nor interesting! Does this have to be true? Can math be engaging, interesting, both? The answer is yes! But, as noted by Dewey, engagement is necessary but not sufficient (1938). In 2018, Darling-Hammond and Cook-Harvey found that learning is social, emotional, and academic. Therefore, we must foster play and nurture joy in learning environments where students feel emotionally and academically safe. We love Math Workshop because it truly engages students in learning. It sets the stage to capture and retain the attention of all students through agency, social interaction, and gamification. To hook students, we invite and consider what they are interested in, intrigued by, or curious about and what makes up their unique identities. Though it sounds simple, maintaining the engagement of all learners throughout the year will challenge you. As you think about this, consider the following four facets of *Design Options for Welcoming Interests & Identities* and answer the questions that follow.

> 1. *Optimize choice and autonomy*: In what ways am I optimizing choice and nurturing autonomous learners?
> 2. *Optimize relevance, value, and authenticity*: How do I connect to students' lives?
> 3. *Nurture joy and play*: In what ways do I nurture joy and play in my classroom?
> 4. *Address biases, threats, and distractions*: Do I address biases, attend to threats, and try to diminish distractions in the classroom? How?

This chapter explores each of these facets and offers a plethora of practical tools and strategies you can use in pursuit of each of these goals. Let's begin.

Optimize Choice and Autonomy

The first facet of welcoming interest and identity—and one of the hallmarks of Math Workshop Plus—is offering student choice and building autonomy. In a Math Workshop Plus classroom, everyone is working toward the grade-level standard, but students have options to select how they work and how they demonstrate their learning. How else can you allow students to make choices? There are six primary areas where you can offer your students choice:

1. How to explore the content, including which tools to use (what students want to do)
2. With whom to work (alone, partner, group)
3. Where to work (workstations, desk, floor, computer)
4. How to practice (draw a picture, write an equation, take a photo of a manipulative, etc.)
5. How students assess and show what they've learned (written explanation, audio recording, photo, etc.)
6. What will be the sequence or timing for completion of tasks

Research shows that when students experience choice they are more engaged and motivated to learn (Assor et al., 2002; Marzano, 2001; Merrill & Gonser, 2021). Marzano (2001) notes that "choice in the classroom has been linked to increases in student effort, task performance, and subsequent learning" (p. 14). Think about yourself. You probably prefer having a choice to not having a choice, but perhaps if there are too many choices, you get overwhelmed. It is the same in your classroom. Students vary in their need for choice. Some want a variety of choices, and others just want a few. As you plan Math Workshop Plus, think about the kinds of choices you'll offer and in what ways the choices are scaffolded.

> For students to thrive and achieve at high levels, they must be interested and emotionally invested in their learning. (Dabrowski & Marshall, 2018, p. 1)

To get you started, let's look at some of the choice factors within the six elements more closely.

Choice in How to Explore the Content

There are many ways to give students choices for how they spend their math time. When considering options, think about what your students need and then reflect on your style of teaching and learning. Some teachers prefer a station rotation model, while others prefer menus. Keep in mind that what works for you may not work for your students, and flexibility really is the key to success and sanity! If choice is new for you, here are some ways to offer choice that are simple:

1. Problem-solving: students choose their approach/strategy
2. Workstation playlists: students select options from a menu

Problem-Solving Approaches

A problem-solving workstation that is up all year gives students countless opportunities to experiment with different ways to solve problems. It also normalizes problem-solving and gives students lots of practice in a low-risk environment. As you explore strategies during whole-group instruction, memorialize them on anchor charts so students will know how to approach problems on their own in the workstations. Offering different types of practice keeps students engaged. For example, What's the Question and Picture Problems are two structures that allow students to think divergently (see Tables 2.1, 2.2, and 2.3).

Table 2.1 • Kindergarten Example

Choose from the following activities.

Structure	Instructions	Example
What's the Question?	Students look at a model and tell a story that matches the model.	The answer is 5 dogs. What was the story?
Picture Problems	Students look at the picture and make up a story about what they see.	This would be a bin of pictures that students could choose from. They might choose a picture of 2 zebras and 5 giraffes. They would then have to tell and write a story about this picture, using a rubric.

Table 2.2 • Third-Grade Example

Choose from the following activities.

Structure	Instructions	Example
What's the Question?	(There is a pile of answers.) Pick an answer and write the story. Illustrate the story.	The answer was 12 apples. It was a multiplication problem. What was the question?
Picture Problems	(There is a pile of pictures). Pick a picture and write the story. Model it in a different way.	Show students pictures of 3 pancakes on 4 plates. Ask the students to tell and write a multiplication and a division story.

Table 2.3 • Sixth-Grade Example

Choose from the following activities.

Structure	Instructions	Example
What's the Question?	(There is a pile of answers). Students pick an answer and write a story.	The ratio was 3 to 5. What was the question?
Picture Problems	(There is a pile of pictures). Students select a picture from a box of pictures and write a math story about it.	Show a picture of clothes in a sales brochure. Ask students to make up problems about buying clothes on sale.

Workstation Playlists

When you use a playlist rather than a station rotation model, students choose which activities they want to engage in while working in workstations for practice. During workstation time, you might offer options to practice specific skills through card games, dice games, board games, computer applications, or projects. When combined with goal setting, students are positioned to make purposeful choices that target the skills they need to learn. Using a menu, choice board, or playlist keeps students organized and provides the right level of structure so the choices are not totally open-ended or overwhelming. See Exhibits 2.1, 2.2, and 2.3 for examples.

Exhibit 2.1 • Choice in Workstations—Primary Example

Working on Doubles

Goal: I can recognize, solve, and recreate doubles problems.

Practice Choices

- ☐ Card Game—Pull a card and double the number
- ☐ Board Game—Doubles
- ☐ Dice Game—Roll and sort facts, doubles, and others
- ☐ Domino Game—Sort Doubles

Exhibit 2.2 • Choice in Workstations—Upper Elementary Example

Practicing Fractions

Goal: I can add fractions with like denominators.

Practice Choices

- ☐ Playdough Fractions
- ☐ Finding fractions of the set
- ☐ Card game
- ☐ Digital game

Exhibit 2.3 • Choice in Workstations—Middle School Example

Integers

Goal: I can add and subtract integers.

Practice Choices

- ☐ Make a poster about integers in real life.
- ☐ Make a social media poster (to post on our classroom Pinterest board or explain on Tik Tok) about adding or subtracting integers.
- ☐ Make up a board game or card game to practice adding or subtracting integers.

Choice With Whom to Work

Sometimes students have the opportunity to choose with whom they want to work. For example, when practicing in a workstation, they could either play a game alone, with partners, or in small groups. If working alone, students might play a game by themselves where they can practice a skill and check their own answers. In partners, students can play dice, dominoes, or a card or board game together. Partners are a great way for students to think about, reflect, and discuss what they are doing. Playing similar games in small groups has its own energy and benefits. Group games teach students how to be patient, take turns, and help each other. We tend to do more small-group games in the Guided Math group.

An additional bonus to having students work alone, with partners, or with groups using dice, dominoes, cards, or board games is to collect formative assessment data and help students formulate goals about the skills they want or need to practice more.

Choice About Where to Work

Choice doesn't have to be complicated or require lots of planning. Something as simple as giving students the option to work in different spaces is an easy way to get started. For example, students may choose to work on the floor, at tables, on vertical whiteboards, or using computers. We've seen students sitting on top of desks, under desks, and on bean bag chairs. Whatever you can tolerate and doesn't present a safety issue works. If that seems like too much, students could also choose the order in which they complete tasks, such as working on the computer before playing a game or vice versa.

Choice and Autonomy About What, Why, and How to Practice

Part of optimizing choice and autonomy is about students reflecting not only on *what* they are choosing but also on *why* they are choosing it. Taking the time to reflect with students about their data—and making a goal and plan based on that data—sets them up for making successful and impactful choices. What does the student need to practice? Why? What should that practice look like? In what ways do they want to practice it? This also could extend to homework menus, where they decide based on their learning goals what they need to practice and how they are going to do it. We will discuss this further in subsequent chapters, but students should understand the connections between these things.

If you are thinking through the lens of math proficiency, great, but also consider the whole child. Goals can also include

- getting started right away;
- listening to and being respectful of peers; and
- being a gracious loser.

The point is, students need opportunities to reflect on their choices, discover which modalities work best for them, and be accountable for how they spend their time.

One way to do this is through a menu that includes "must-do" and "may-do" items. As seen in Exhibits 2.4, 2.5, and 2.6, students have "Must Dos" and "May Dos" for the week or whatever amount of time works for you. The way you approach this can vary, but many teachers have one Must Do each day, which typically involves practicing what was learned in the mini lesson. If using a program, this could be the workbook page, depending on how long it takes to complete. If everyone does the Must Do at the same time, you can move around the room and see who has it and who needs more instruction. Once the Must Do is complete, students have autonomy to make choices from the May Dos. This structure also gives students choice in not only how to practice but also choice about how to show what they have learned. The artifacts serve as proof of various activities—for example, a recording sheet. Most of the activities have a self-checking element so that students can self-assess. This could look like an answer key to check work against once everything is done. This could also look like rubrics and checklists. This Must Do/May Do structure also gives students the opportunity to choose the sequence and/or timing for completion of tasks. They learn to pace themselves, so they meet the expectations within the time allotted, which is something that will serve them well beyond the math classroom!

Exhibit 2.4 • Must Dos and May Dos—Primary Example

Math Workshop

Instructions: Complete your Must Dos. Then choose which May Do activities to do.

Must Do	May Do
Fluency Folder every day	Make your own board game
Must choose a problem-solving activity	Count around the room
	More Pumpkin Math
Place Value math activity—1, 2, or 3	Spider Math
	Fact Concentration
Guided Math group visit 2x	Pumpkin Math

Exhibit 2.5 • Must Dos and May Dos—Upper Elementary Example

All About Multiplication

Must Do	Can Do With a Friend	Can Do by Yourself
Meet with teacher	Make a video about arrays or equal groups	Make your own board game
Review game with a friend	Make an array game	Make a poster
Fluency Journey practice from your folder	Play Circle and Stars	Make a social media post
Friday Quiz	Free choice	Create a card game/ flashcards to practice

Exhibit 2.6 • Must Dos and May Dos—Middle School Example

All About Decimals		
Must Do	**Can Do With a Friend**	**Can Do by Yourself**
Meet with teacher	Make a video	Make your own game
Review game with a friend	Make a poster about using tools and models	Write a blog post or poster about decimals in real life
Fluency Review	Play a card or board game	Make a social media post about different tools to learn about decimals
Friday Quiz	Free choice	Create a Decimal Quiz with the answer key

The Must Do, May Do protocol is powerful for allowing choice while maintaining accountability. Students are held accountable to complete their work and to document their choices and thinking through accountability sheets, journal entries, or in a folder that holds the work for the week.

Regardless of what options are available for capturing their work—on a laptop, paper and pencil, or dry erase boards—remember to share and model examples of evidence in any modality. For example, if students work on dry erase boards instead of on paper, how can they show what they've done? You might want them to take a picture using a device and share it to Google Classroom or add it to their digital journal in Seesaw. If students keep a physical journal or a folder of work, share exemplars for what the entries should look like. Posting these on anchor charts, or offering templates, helps to ensure that this happens efficiently, effectively, and completely. Taking time to check in with students throughout the week allows you to monitor their progress and offer feedback about the quality and content of their artifacts. The recipe for success is clear expectations, accountability, and just-right options in workstations.

Connections to SEL

Reflecting on Self-Management

Learning opportunities such as those discussed in the section on choice and autonomy help students own their learning, set and achieve goals, and create pathways of learning that they helped to design. Students should have regular opportunities to reflect on how they practiced self-management. Including a quick self-assessment on an exit ticket helps to build accountability and helps students set a plan for improvement next time. In this way, students become self-motivated.

Student-Made Board Games

One workstation option we love is allowing students to make their own games. This type of choice incorporates creativity, reflection, and the application of skills and knowledge. If you haven't done this before, you can create a simple "game proposal" form and have students pitch their idea(s) to you. Their proposal should include what they know about the topic, what the game will involve, what they are hoping to learn/practice from it, and a timeline to complete making the game itself. Once you approve their proposal, they can use an agreed-upon amount of time each day during Math Workshop Plus to work on making the game. Once complete, they can use their game to practice the targeted skill. Once they've mastered the skill, they can add their game to the classroom collection to be played by other students. As seen in Exhibit 2.7 and 2.8, examples include, but are not limited to, traditional game boards, Tic Tac Toe, Four in a Row, and more. You might also use these templates to create individualized practice according to student and small-group needs.

Exhibit 2.7 • Tic Tac Toe Board

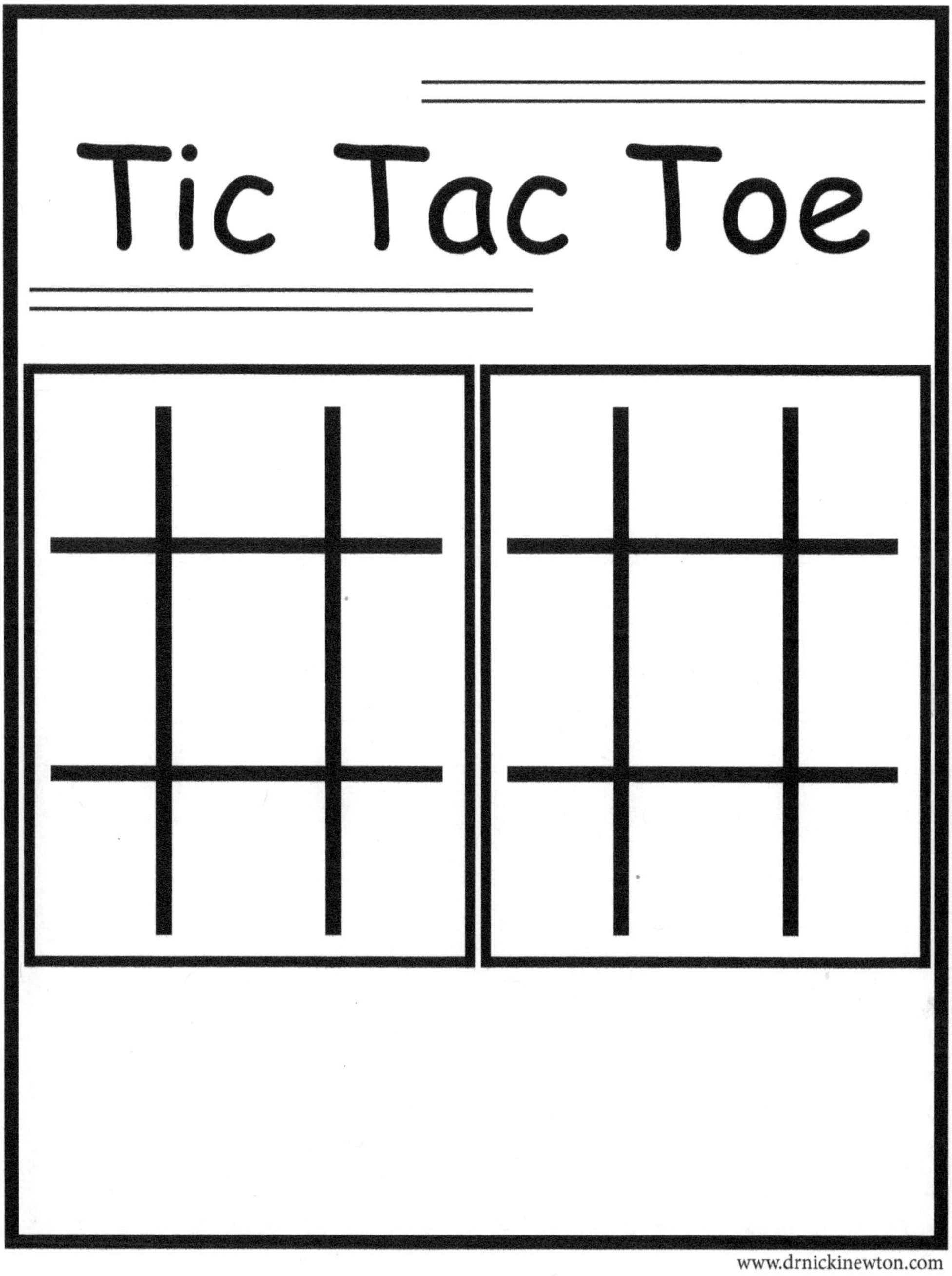

Source: Dr. Nicki Newton

 Download a blank Tic-Tac-Toe Board at https://companion.corwin.com/courses/ MathWorkshopPlus

Chapter 2 • Design Options for Welcoming Interests and Identities

Exhibit 2.8 • Four in a Row Board

FOUR-IN-A-ROW

Take turns. Pick a card and cover up the problem that makes that difference. Whoever gets 4 in a row first wins! Or, just play cover up with a partner, by picking cards and trying to cover up a row or column.

www.mathfactfluencyplayground.com

Connections to SEL

Self-Motivation, Self Reflection, Self-Efficacy, and Relationship Building

When students have opportunities to create their own games based on what they need to practice, they become self-motivated, better manage their learning, and are poised to set and achieve their goals. They learn how to accurately assess their strengths, understand areas where they need practice, and how to manage their time and resources. By making their own games to practice based on self-reflection, students build self-efficacy.

As an added benefit, students work on relationship skills by playing, negotiating, listening, and compromising with others. Throughout Math Workshop Plus, students thrive in a caring environment where they learn in a community and feel a sense of belonging and connectedness with others. Students continuously build relationship skills through working with diverse individuals and groups, working cooperatively, depending on each other for help, communicating clearly, and resolving conflicts when needed.

Equity Check

Learning That Is Varied, Authentic, and Connected to Students' Lives

Giving students choice in the classroom is central to promoting equity and mathematics. We design Math Workshop Plus in a way that considers the many different factors "that influence individual variation in motivation including culture, neurology, personal relevance, and prior knowledge" (CAST, 2024). One goal of equity is for all students to have the opportunity to learn in meaningful and engaging ways. It falls to us to tap into interests and incorporate identities into the very fabric of school life. Another goal of equity is for all students to attain academic success. That comes when learning is authentic and connected to the lives of students. It means that we individually scaffold students into success by knowing who they are and what they need to succeed (Intercultural Development Research Association [IDRA], 2020).

Optimize Relevance, Value, and Authenticity

As mentioned previously, relevance promotes engagement, grounds thinking in context, and increases retention. When you intentionally offer authentic math activities that add value to students both in and out of school, learning becomes relevant. This doesn't have to be hard or require a ton of prep. It can be as simple as focusing on routines and practice based on the daily realities of students—for example, engaging in a problem-solving routine based on typical after-school routines such as stopping at the store, babysitting, cooking, working, doing chores. Riddles and Number Talks can also be contextualized around the people and places that students know or are interested in, such as celebrities or sports figures.

Integrating Routines

As you know, Math Workshop starts with a routine to get the math juices flowing! Routines such as I Was Walking Down the Hall and Number Riddles serve a variety of purposes in a universally designed Math Workshop. Launching the workshop with a routine activates thinking and generates excitement. In addition, routines can foster equity and community because they are typically low floor–high ceiling. This means that all students have entry points and have something relevant to contribute regardless of their math knowledge or current mathematical proficiency.

I Was Walking Down the Hall

In this routine, students get to "get in, where they fit in." Everybody has a way to participate. The questions are open, and everybody has an opportunity to contribute (see Exhibit 2.9).

Exhibit 2.9 • Example of I Was Walking Down the Hall

Teacher:	I was walking down the hall, and I heard Jamal say 12. I wondered what was the question?
Student A:	Was it 6 + 6?
Student B:	Was it 24 − 12?
Student C:	Was 3 × 4?
Student D:	Was it 24 ÷ 2?

Each contribution is worthy and valid. If a student gives an incorrect response, you can use questioning and add scaffolds to support success. As an added benefit, this routine encourages students to actively and respectfully listen to their peers while they wait for their opportunity to share, building the muscle of self-regulation (CASEL, 2020).

The best thing about the routines we share is that they get students thinking, increase discourse, build confidence, and develop a sense of community within the classroom. Because this is so important, you will hear this theme many times throughout the book.

Number Riddles

What's This Number?

Researchers have found that number lines are really important (Institute of Education Sciences [IES], 2021). Students need to understand number lines throughout grades K–12. Doing a routine where students have to work with number lines every week is necessary to build number sense. We have to think about ways that all students can begin to grasp this concept. In the primary grades (K–1), students should be working mainly with number paths (Brownell, 2023; see Exhibit 2.10, Figure 2.1).

In the following exhibits, the students are asking the questions. This is important. The teacher has modeled the activity with students many times and eventually gets the kindergarteners to ask the questions using the correct vocabulary.

Exhibit 2.10 • Counting and Cardinality

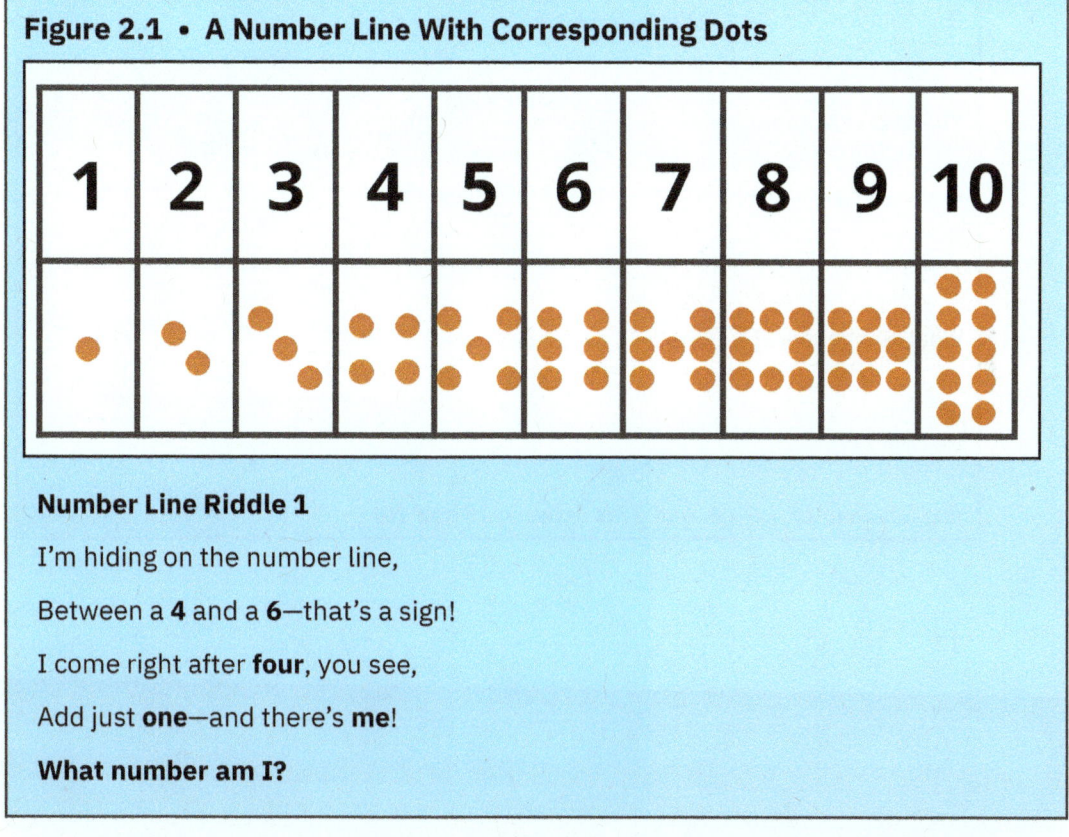

Figure 2.1 • A Number Line With Corresponding Dots

Number Line Riddle 1

I'm hiding on the number line,

Between a **4** and a **6**—that's a sign!

I come right after **four**, you see,

Add just **one**—and there's **me**!

What number am I?

There are so many different number line activities, including number line riddles. What is important about these energizers is that the students all have number lines and/or visuals in front of them so that they can act the problems out (see Exhibits 2.11 and 2.12).

Exhibit 2.11 • Addition and Subtraction Number Line Riddles

The Jumping Frog

A frog jumps 3 spaces forward on a number line.

It starts at 7 and lands on a number.

Then it jumps 4 spaces back.

Where does it land?

Answer: 8

(Explanation: 7 + 3 = 10, then 10 − 4 = 6)

The Secret Number

I'm a number on the line, greater than 15 but less than 20.

If you subtract 8 from me, you get 9.

What number am I?

Answer: 17

(Explanation: 17 − 8 = 9, and 17 is between 15 and 20)

Exhibit 2.12 • Fraction Number Line Riddles

The Hidden Fraction

I'm a fraction on the number line between $\frac{1}{2}$ and 1.

If you subtract $\frac{1}{4}$ from me, you get $\frac{1}{2}$.

What fraction am I?

Answer: $\frac{3}{4}$

$\left(\text{Explanation: } \frac{3}{4} - \frac{1}{4} = \frac{1}{2}\right)$

$\frac{1}{2}$		$\frac{1}{2}$	
$\frac{1}{4}$	$\frac{1}{4}$	$\frac{1}{4}$	$\frac{1}{4}$

The Fraction Challenge

Start at $\frac{2}{3}$ on the number line.

Move $\frac{1}{6}$ spaces forward.

Then move $\frac{1}{2}$ spaces back.

Where do you land?

Answer: $\frac{1}{3}$

$\left(\text{Explanation: } \frac{2}{3} + \frac{1}{6} = \frac{5}{6}, \text{ then } \frac{5}{6} - \frac{1}{2} = \frac{5}{6} - \frac{3}{6} = \frac{2}{6} = \frac{1}{3}\right)$

$\frac{1}{3}$		$\frac{1}{3}$		$\frac{1}{3}$	
$\frac{1}{6}$	$\frac{1}{6}$	$\frac{1}{6}$	$\frac{1}{6}$	$\frac{1}{6}$	$\frac{1}{6}$

Increasing Relevance

If you launch the workshop with a routine that capitalizes on relevance, connections to the real world don't need to stop there. Through Math Workshop Plus, you can get students thinking about and connecting math to real life. The learning experiences you design should make students active participants in their learning, allowing them to explore concepts up close, and experiment with ideas (CAST, 2024). This will illuminate how and why their learning is relevant to their daily lives and showcase that there is value in learning how to do math.

You can capitalize on authentic situations where math occurs in your classroom, at home, or anywhere it is visible to students. This may mean tweaking the contexts presented in your curricular materials or creating your own contexts that resonate with students and experiences they can relate to. A kindergarten teacher we know takes her students on a field trip through the school to find math in action. The lunch ladies love showing how they measure, count, and cook things at different temperatures for different amounts of time. Think of how much math is right in your school!

The mini lesson gives you another great chance to connect what students are learning to what they know, see, and do in their daily lives. Over the years, we've missed so many opportunities to make "personalized and contextualized to students" lives (CAST, 2024), which makes learning more powerful. In recent years, we shifted to proactively curating these opportunities, and not only is it effective, it's also more fun!

For instance, when studying decimals, we start by asking where students have noticed decimals in the real world. Then we share menus from restaurants and fast-food places they may visit. Since they are likely familiar with the menus and understand the context, the engagement is natural. They can reason about the math because it makes sense to them. They are interested because it feels nonacademic and meaningful. When we think through the lens of UDL, we add visuals from the websites, integrate videos from commercials, and get creative with making up orders and generating problem-solving scenarios involving the menu items. This makes math fun, meaningful, and culturally relevant.

When thinking about cultural and social relevance, students' sociocultural realities should be woven throughout activities in Math Workshop Plus (Banks & Banks, 2007; CAST, 2024; Ladson-Billings, 1995). That may sound laborious or unrealistic, but you may already do things such as integrating the names of students, the places they go, and the things they do into the problems you present. Recently, we used AI to generate problems focused on percentage increase in the context of concert ticket sales of a popular artist, and the students were all in! Tapping into the world students live in helps them to make connections, fosters sense-making, and makes math seem less like an exercise in answer getting and more like a valuable life skill.

Increasing Value

If you asked your students what value their math lessons add to their lives, what would they say? What about the tasks they complete? Do you think they would say they add value to their lives or the lives of others?

Students are more likely to take ownership of the knowledge they are learning when they find value in the process and can share it with others. In other words, when there is an authentic audience, the game changes! In the 1990's, Alison had students apply their math skills to complete a design and architecture project which culminated with a mock pitch to the Town Council to get their plans "approved." There was so much math involved in the project, including budgets, geometric concepts, and calculations. In retrospect, it was project-based learning, though it wasn't called that back then. Students worked so hard, and when the day came to present their work, they dressed up like professionals and took it so seriously!

We see students raise their own stakes and approach problems and projects in a very different way when there is perceived value to the outcome. Recently, a group of fourth graders was tasked with visiting third-grade classrooms to present the topic of multiplication. The students were extremely vigilant about their work and very attentive to what they said and showed. They were motivated to create a good presentation, not only for the teacher, but for the third graders. The knowledge that what they were doing mattered and added real value to the younger students motivated them.

We see students raise their own stakes and approach problems and projects in a very different way when there is perceived value to the outcome.

Increasing Authenticity

Since students are interested in math that connects to real life, we love using 3-Act Tasks. If you aren't familiar with them, 3-Act Tasks use multimedia such as videos and photos to pose real questions about things students can relate to and be curious about. They are framed in such a way that only a little information is presented so students formulate the question and seek the additional information needed to solve it. There is a sizable collection of these available for free online. Although they are typically aligned with a grade-level standard, we have found them to be versatile and may be used in different grades for different purposes.

For example, there is a 3-Act Task called Sugar by Dan Meyer (n.d.). Although it is written as a ratio problem, it works great in fourth and fifth grade as a division problem. In this task, students watch a video of a man sitting at a counter and eating packs of sugar. Sitting to his left and right, there are two people drinking sodas and looking at him, appalled by his behavior. The task is all about exploring how much sugar is in a soda, and it turns out to be a public service announcement about consuming healthy drinks. Students love the task and find it very engaging and relevant. The follow-up assignment is to go home and figure out how much sugar is in their favorite foods, which leads to some great conversations!

Connections to SEL

Leaning Into Others' Perspectives

When interweaving relevant and authentic activities throughout Math Workshop Plus, we are building social awareness. Students get to learn about diverse communities, participate in perspective taking, and recognize, understand, and celebrate diversity in our world. Think about how you create opportunities for students to lean into others' perspectives with curiosity. We must provide opportunities to recognize and acknowledge the inherent strengths in others.

Equity Check

Authentic Connections Support Equity

Making lessons relevant to students' authentic lives and daily experiences is not just helpful for engagement and enduring understanding but is a critical element for promoting equity in the mathematics classroom. Students can more effectively access lessons that match their contexts and experiences. They are more likely to see the value in lessons that pertain to them. It is essential that we bring in the cultures of students when planning lessons. Everybody's culture, language, food, and ways of being should be not only valued and discussed but highlighted and integrated into classroom activities when appropriate.

Nurture Joy and Play

Research shows how important the notion of joy is for learning (Hough, 2022). "The truth is that when we scrub joy and comfort from the classroom, we distance our students from effective information processing and long-term memory storage" (Willis, 2007, para. 2). Wow! Think of yourself as a learner, and you will probably agree. When students are happy and engaged in class, it jump-starts learning. Willis notes that when engaged in joyful learning, students "achieve higher levels of cognition, make connections, and experience 'aha' moments" (para. 3). Math Workshop Plus is a space where students can laugh and play and have fun while they are learning.

You will use Math Workshop Plus to intentionally plan for a purposeful, engaging, standards-based, academically rigorous learning environment. Throughout their math journey, students will experience the ups and downs of learning. Math Workshop Plus positions learners to experience the struggle of persevering, learn the power or collaboration and discussion, and feel the triumph of finally getting it. Irby notes that "joy at school and in learning is a foundation from which students gain the confidence that academic struggle is temporary and worthwhile" (cited in Hough, 2022).

"Research shows that play has enormous learning and social benefits across all ages" (Mader, 2022, subtitle). You can set a playful tone when launching energizers and routines in the beginning of Math Workshop, including things such as drum rolls and making things feel like a game show. As students transition to workstations, they get to play different types of games, and you may even opt to play a game with students during your small-group time. Whether you teach kindergarten or eighth grade, play is not fluff when it is used to target specific skills, and the benefits go beyond getting better at math. Play supports the SEL competencies in countless ways and elevates the social aspects of learning. Play has a way of lowering the affective filter so that students can engage with the math without being self-conscious. Play allows students to take risks in ways that they would ordinarily be more hesitant to take. Play is a universal right of all children (United Nations, 1989, Article 31).

Connections to SEL

Joy and Play Foster Harmonious Relationships

Joy and play are based on relationship skills and social awareness. Students learn to think about others in relationship to themselves. They learn to agree, respectfully disagree, negotiate, and compromise so that they can harmoniously live and learn in the classroom community. Relationship skills are an essential ingredient in the success of your Math Workshop Plus and should be supported with anchor charts, examples, nonexamples, and ongoing reflection.

> **Equity Check**
>
> Inclusivity
>
> Equity as inclusivity is interwoven throughout concepts of joy and play. We have to "create space for learners to find joy through connections to their identities, sense of self, and communities" (CAST, 2024). Through Math Workshop Plus, we build inclusive environments that welcome diverse identities and foster a sense of belongingness, and this entails having "difficult conversations" when the need arises. As Paley said, "You can't say, you can't play" (Paley, 1993). In this book, she talks about how we must stop and discuss how children are treating each other when they are playing. When needed, we must interrupt instances of prejudice and mistreatment, based on whatever difference. Sometimes students will exclude others from play, based on socioeconomic status (SES), race, language, ability differences, and other things. One of the six goals of educational equity is about full participation and integration of all students in all programs and activities (IDRA, 2020). We must rise to the equity challenge and create environments that stand up for all students and name and rectify injustices.

Address Biases, Threats, and Distractions

Your math classroom should be a place that is bustling, safe, noisy, and well-organized. Students should be talking, moving, playing, discussing, working, reflecting, creating, and learning. To foster a learning environment that truly values all learners, strategies to confront and eliminate biases are part of your planning. In a Math Workshop Plus classroom, the social and physical environment are vital and directly impact teaching and learning (Bronfenbrenner, 2005; Dewey, 1934; Erikson, 1963; Montessori, 1967; Piaget, 1959; Vygotsky, 1978). The way you run Math Workshop Plus reflects your theoretical and ideological beliefs about teaching and learning (Read, 2019). We want students to see themselves as kind, compassionate, helpful, brilliant, productive, creative citizens of the world, and we want to support you in creating a space where all students experience deep joy and a sense of wonder as they learn math. What can get in the way of this? Understanding and recognizing bias and threats can help to proactively mitigate them.

We have found that one way of addressing biases is to actively cultivate an environment where everyone feels like they belong. Students need to own the classroom. You play a huge role in creating a culture that communicates that everyone matters, all are in this together, and it is "our classroom." Be aware of how bias can creep in when you address conflict or utilize boxed curriculum.

In a universally designed Math Workshop, making an explicit plan for distractions will pay off. As you plan, consider your level of clarity in every area. Do students have a clearly articulated schedule? Do they understand how much time to spend at each station, when to move on, and how to transition without disrupting others? Math Workshop Plus runs on systems that help students know what is going to happen and when it is going to happen. Predictability and consistency are the keys to making it work. This may look like posting schedules, anchor charts, and timers, which all help add structure. That said, when you keep learner variability in the forefront of your mind, you may have to remind yourself that flexibility in some areas will also help minimize distraction.

> Math Workshop Plus challenges you to view your students through the lens of Universal Design to ensure equity and access for all.
>
> - Are your signals verbal, visible, or digital?
> - What do your students need?
> - Are the timers visible so students can pace themselves during independent workstations?

One area where variability is obvious is attention span. In recent years, we've seen attention spans shrink, but as a rule, researchers and psychologists note that students should be able to focus for the duration of their age plus a few minutes. That said, you may find some students who can effortlessly attend for up to double their age in minutes (Kannas et al., 2010; Schmitt, 2024; Ward, 2020). If you teach kindergarten, this means you may be able to hold your student's attention for about six or seven minutes, whereas if you teach fifth grade, it could be more like 20 minutes.

Keep this in mind as you design your workstations. Menus are especially helpful to match the variability of attention spans because students determine their own pace. This is important for small groups as well. Not all students need the same number of minutes, even if they're in the same group. Don't be afraid to be flexible! If a student is ready to leave the group and move to other work, release them. For some students, this will have to be more scaffolded than for others, and we have learned that, to limit potential disruptions to other students, you should have a clear list of options that students may work on independently if they find themselves without a partner or with a few extra minutes.

As you continue to think about addressing threats and distractions, think about your physical classroom space. Is everybody able to maneuver and find a place that is comfortable for their body? The "optimally designed classroom space offers children a setting for exploration, reflection and learning" (Read, 2019, abstract) by the nature of the way learning opportunities are created. Light, seating, color, noise levels and more all contribute to making the environment conducive for learning. In Math Workshop Plus, the teaching and learning experiences are created, fostered, and promoted by your classroom's design elements and principles. These include movement around the classroom; the layout of desks, chairs, and seating spaces; the color of the classroom walls; and the lighting within the class (Read, 2019). We have personally been in rooms where the lighting is intentionally low to foster a calm aesthetic, but we couldn't see anything!

> Children's experiences in the designed environment are impacted by a classroom's myriad design elements and principles, which include the circulation, spatial layout, degree of enclosure, material, color, and lighting of the space (Read, 2019, para. 2).

Noise Levels

Let's talk about noise! This is one area where we see extreme variability among teachers, and we want you to think about whether your norms are more about *you* or about your students. We see classrooms where teachers love to have music on. Whether classical, pop, or something else, it's a constant stream of sensory input that doesn't work for everyone. We suggest discussing noise with your class and coming to consensus regarding what works for everyone and then holding students (and yourself) accountable.

In Math Workshop Plus classrooms, we find noise meters to be extremely helpful. You can use old-school noise meters (paper and pencil) or new school ones (digital). Through the lens of Universal Design, we consider how noise levels impact different students. For example, students with sensory sensitivities might be negatively impacted by loud classrooms or be distracted if computer sounds are on in the background. One option is to work with occupational therapists to determine if headphones or noise-canceling earbuds might help during workstation time when different things are happening at the same time. You might find that some students opt for headphones because the "daily hum" gets to be too much for them (Kluth, 2020, p. 125). Once again, having options and choice will allow students to make the adjustments they need and empower them to advocate for the conditions that work best for their learning.

Timed Tests

It may seem random to see timed tests pop up here. Why do you think timed tests are included in a discussion about minimizing bias, threats, and distractions? Fluency with basic math facts is surely a goal in every district, school, and classroom, and we know that fluency with basic facts is a key outcome for all students. Unfortunately, for many years the definition of fluency has been tied to rote memorization, and assessment has often taken the form of speed drills. Despite decades of research indicating that timed tests are less reliable, less valid, less inclusive, and less equitable (Gernsbacher et al., 2020), their use persists. This practice undermines your ability to limit threats and create a safe and inclusive environment. To walk the talk of respecting differences, you can't have an end-all, be-all metric like timed tests. These tests have been linked to math anxiety, suggesting a negative impact on student confidence. They have also been shown to increase stress, decrease motivation, and in some cases, create fear of mathematics as a whole (Boaler, 2014).

In your universally designed Math Workshop, students can achieve fluency through a wide variety of games using dice, cards, spinners, or digital tools. Proficiency may be demonstrated orally, in writing, or using technology, but not through a timed test. Of course, we favor *Math Running Records* (Newton, 2016) as a way to holistically assess both fluency and math disposition. This approach minimizes threats, fosters a growth mindset, and values strategic reasoning.

Connections to SEL

Building Self-Awareness Helps Combat Threatening Environments, Biases, and Distractions

Addressing threatening environments and biases isn't just the work of the teacher, but the work of all those in the classroom, including students. Math Workshop Plus encourages students to be self-aware. You want all students to be able to label their feelings, think about how those feelings are impacting their behavior, and realize how their behavior is impacting others in the classroom community. Encourage responsible decision-making through the choices you offer and support self-reflection through the debrief. When needed, use books or role-play to build capacity when you see lagging skills. This is important because in an equitable environment biases and threats must be addressed.

Equity Check

Creating Opportunities for All Students to Feel Safe and at Peace in Your Classroom

Addressing biases, threats, and distractions is another way to promote equity. This means ensuring all students have equal opportunities to live peacefully and safely in your classroom, learn from you and each other, and engage in varied activities throughout the workshop (CAST, 2024). Equitable systems for addressing what happens when things go wrong in the classroom enables a sense of belonging, safety, and justice for all students. A goal of educational equity is that all students participate in educational "pursuits of learning and excellence without fear of threat, humiliation, danger or disregard" (IDRA, 2020). Students need to know and believe that everybody is treated the same and that there are no teacher favorites.

Say Bye-Bye to Barriers When You . . .

- Give authentic choices
- Promote autonomy
- Do relevant and valuable activities
- Offer authentic learning experiences
- Nurture joy and happiness
- Build a classroom community that is safe and predictable
- Cultivate a sense of belongingness

Action Plan

Assess your current practice in Designing Options for Welcoming Interest & Identities. Download the Chapter 2 Self-Assessment at https://companion.corwin.com/courses/MathWorkshopPlus.

Try It! Where Do Students Have Choice?

◇◇

Identify three areas where you can allow students to have choice. Try one that seems the most realistic for you and ask students for their feedback.

 Connections

◇◇

Share something you have done to recruit and sustain interest with your colleagues or on social media using #mathworkshopPLUS.

Chapter 3

Design Options for Sustaining Effort and Persistence

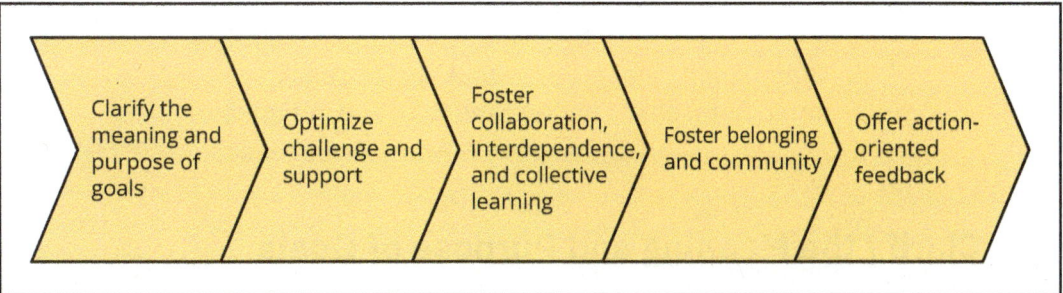

Throughout Math Workshop Plus, you will constantly think about ways to encourage effort and develop persistence among students. Your goal is to get students to reflect on the level of effort they are putting into their work and to cultivate their intrinsic motivation to persevere so effort is sustained throughout the workshop. To sustain effort is to give the physical and/or mental activity to make something happen to achieve a goal (Inzlicht et al., 2018).

Think about how you will create opportunities to set learning goals and how students will understand their importance. Consider how you will intentionally offer challenging opportunities to explore mathematics while fostering a community of engaged, collegial, helpful learners. Think about current and future avenues to give immediate, helpful, action-oriented feedback that helps students to achieve their goals.

As you consider the ways you can help students in *Design Options for Sustaining Effort & Persistence* throughout Math Workshop Plus, use these questions to guide your thinking:

1. *Clarify the meaning and purpose of goals*: Is mathematical goal-setting an integral part of our classroom culture?
2. *Optimize challenge and support*: Are there plenty of opportunities for students to reflect on how they are challenging themselves and what supports they might need?
3. *Foster collaboration, interdependence, and collective learning*: How do we use math routines, workstations, and whole-group activities to build a collaborative learning environment?
4. *Foster belonging and community*: In what ways am I fostering a sense of belonging and community throughout Math Workshop Plus throughout the year?
5. *Offer action-oriented feedback*: Is feedback immediate, easily understandable, and actionable?

Clarify the Meaning and Purpose of Goals

In a Math Workshop Plus classroom, goal setting is an integral part of the student experience. Students need to understand what they are working toward. They should constantly ask themselves, What do I need to learn? What do I already know? How will I go about achieving the goals I set? Notice the phrase "goals *I* set." Simply posting a learning objective that students are expected to obtain causes them to feel separated from the goal. They view it as *your* goal or *the school's* goal for them. They have no ownership in it, and their motivation then is about pleasing you or complying with your goals. It is only extrinsic. In a Math Workshop Plus classroom, a vital objective of goal setting is for students to internalize a love of learning and to cherish the process of learning in and of itself, rather than focusing on external validation (grades, compliance, or pleasing the teacher) or competition with classmates. This is a key part of the culture of learning that is fostered through Math Workshop Plus.

If setting goals in partnership with students is not currently part of your practice, how might this look? You might start by discussing the purpose of goal setting. From there, you can support students by setting up opportunities to set, share, display, and work toward simple goals, co-creating and posting goal-setting anchor charts, posters, or agendas and pausing to recognize and celebrate when goals are achieved. Students benefit from such visual goal-setting tools that help them work toward their goals step-by-step, so they can sustain effort on their journey toward the goal. Let's talk about some ways to set, monitor, track, and celebrate goals.

Set Goals That Inspire Confidence and Ownership of Learning

One of our favorite things about Math Workshop is that it creates an expectation that students own their learning. This shifts our role from "sage on the stage" to "guide on the side." In this capacity, we help students find their strengths and understand where they are on the learning continuum. This positions students to set goals and monitor their progress.

We are going to zoom in and consider how the use of small, attainable goals that are personalized by and for specific students can inspire confidence and increase ownership. Success breeds confidence and motivation, so selecting goals that are realistic, easy to support, and have easily measurable checkpoints set the stage for success and keep students motivated.

Let's look at an example that affects every grade: fluency with basic math facts. Depending on the grade you teach, you may or may not have mastery objectives for basic facts. These are outlined in Table 3.1 for your reference (although most states use these, a few differ slightly).

Table 3.1 • Basic Fact Fluency Benchmarks

Grade Level	Mastery Expectation
Kindergarten	Addition and subtraction within 5
Grade 1	Addition and subtraction within 10
Grade 2	Addition and subtraction within 20
Grade 3	Multiplication and division through 9×9
Grade 4	Multiplication and division within through 12×12

Fluency with basic facts is a perfect area to get started with individual goal setting. Since students acquire fluency at different rates, setting personal goals makes sense. Whether you are teaching in a grade that has a basic fact fluency benchmark or you are not, you will likely want to dedicate space within workstations for students to pursue mastery of the benchmarks. Since there is a desired sequence to the facts, it is not a lot of work for you to set students on the correct path once they know where they are starting from.

An Example of Goal Setting Using a Basic Facts Fluency Progression

Let's illustrate our point through a progression of committing basic facts to memory, but we want to be clear that this is merely an example. Goals can be academic, social, or emotional. They can be short-term or long-term. If fluency is not the goal you want to target, fill in the blank with whatever you want to focus on. Individual goals allow us to better differentiate, to account for variability, and to ensure all students get what they need. They also offer ongoing opportunities to celebrate growth, build confidence, and reinforce motivation (Figures 3.1, 3.2, and 3.3).

1. First, distribute a blank multiplication chart as seen in Figure 3.1.

Figure 3.1 • A Blank 12 × 12 Multiplication Chart

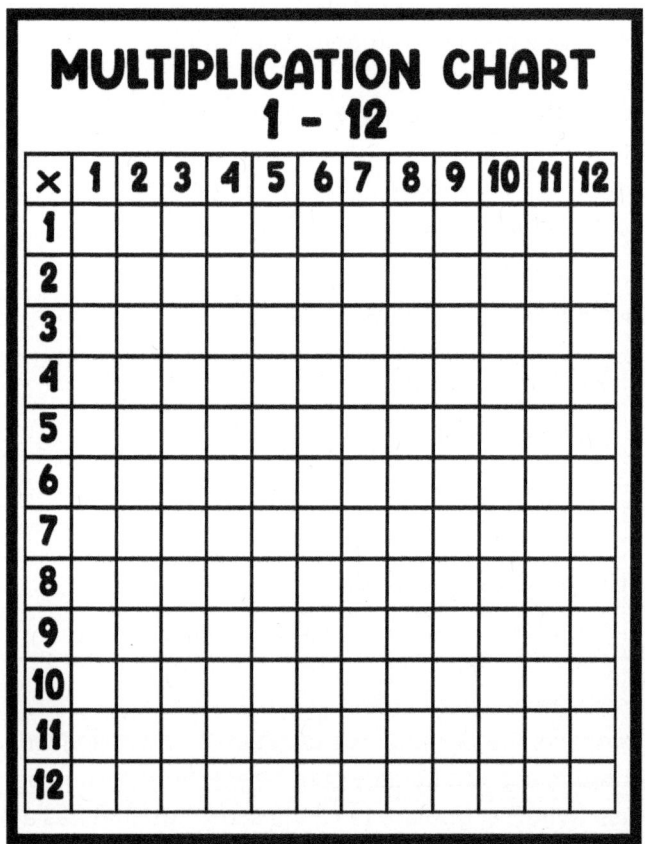

Source: Dr. Nicki Newton

 Download this chart at https://companion.corwin.com/courses/MathWorkshopPlus

2. Next have students fill in the facts that instantly pop into their minds. If they need to figure it out in a way that takes more than 3 seconds, tell them to leave the space blank.

Figure 3.2 • A Partially Filled-In 12 × 12 Multiplication Chart

MULTIPLICATION CHART
1 - 12

×	1	2	3	4	5	6	7	8	9	10	11	12
1	1	2	3	4	5	6	7	8	9	10	11	12
2	2	4	6	8	10	12	14	16	18	20	22	24
3	3	6			15					30		
4	4	8			20					40		
5	5	10	15	20	25	30	35	40	45	50	55	60
6	6	12			30					60		
7	7	14			35					70		
8	8	16			40					80		
9	9	18			45					90		
10	10	20	30	40	50	60	70	80	90	100	110	120
11	11	22	33	44	55	66	77	88	99	110		
12	12	24			60					120		

www.mathfactfluencyplayground.com

Source: Dr. Nicki Newton

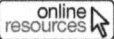

online resources Download this multiplication chart at https://companion.corwin.com/courses/MathWorkshopPlus.

3. Then have students identify where they are based on the facts they know. This is the starting point (see the game board in Figure 3.3), and they will follow the path and set new goals as each set of facts is mastered.

Figure 3.3 • Multiplication Pathways

Source: Dr. Nicki Newton

 Download this Multiplication Game Board at https://companion.corwin.com/courses/MathWorkshopPlus

The power of this practice comes from students selecting their goal(s) (often with your guidance), creating a plan to meet them, and monitoring their progress along the way. Again, we will discuss this in further detail in Chapter 10, but if you are eager to get started, here is a snippet of what this could look like:

Step 1: Identify the starting point.

This works best when you have data. That said, as noted in Chapter 2, we don't support using timed tests to assess fluency. We have shared an idea to get a baseline. For additional ideas check out the books *Fluency Doesn't Just Happen with Addition and Subtraction* (Newton et al., 2019), *Fluency Doesn't Just Happen with Multiplication and Division* (Newton et al., 2024), and/or *Math Running Records in Action* (Newton, 2016).

Step 2: Develop a plan.

Designate time during Math Workshop Plus each week to confer with students about their goals. Take no more than a minute or two to ensure students can answer these questions:

1. What is my goal?
2. Why do I have this goal?
3. How can I reach this goal?
4. About how long should it take to reach it?

Step 3: Monitor progress.

Monitoring progress is done by the student, not you. We recommend you model this and do some role-play so they get the idea. As seen in Chapter 3, offer checklists or other systems to students so they can track their progress.

Connections to SEL

Ensuring That Goals Are Attainable to Maintain Self-Regulation

If students become frustrated, exasperated, or dysregulated when goals aren't met in the anticipated timeframe, reassess the goal. Was it too ambitious for the timeline? Did it miss the mark for the starting point? Do the workstations support the target? Goals should be attainable, so discuss what strategies the student is using and what new strategies may be needed. It may be as simple as changing the modality, leveraging technology, or adjusting the timeline. If students are feeling defeated or unsuccessful, be sure to reinforce the Power of Yet!

As with all aspects of Math Workshop Plus, make this your own. If it feels overwhelming, scale back or do it less frequently. Keep checklists short and simple. Create a quick and easy system for check-ins with students. As mentioned in Chapter 3, take time to celebrate milestones and remember to think about the whole child when developing goals. Table 3.2 shows a few more examples to help you get started.

Table 3.2 • Examples of Student-Owned Goals

Goal: I Will . . .	Tips to Monitor Progress
Get started right away when we transition	Track time on a grid paper graph, chart, or computer
Be a better partner	Use a checklist to reflect
Remember to use my tools and reference folder	Track number of times a week

(Continued)

(Continued)

Goal: I Will . . .	Tips to Monitor Progress
Select a workstation to practice _____ at least 2x/week	Complete artifact/accountability sheet
Show my thinking on exit tickets	Add photos of exit tickets in Seesaw
Use the resources in the Calming Corner when I'm stressed	Track frequency/note what works for them

Connections to SEL

Taking Ownership of Learning

When students take ownership of their learning through goal setting, they build self-awareness. Students must reflect on where they are and what they are doing. They must set goals, work to meet them, and set new goals. Sometimes goal setting might entail working with a goal partner, which requires relationship skills. Students must know how to work with each other, how to give positive feedback, and how to receive feedback. To support this, plan to explicitly teach students how to gracefully give and receive helpful feedback.

Recognize and Celebrate Achieved Goals

Recognizing and celebrating goals individually, as a class community, and with families, sustains momentum and builds positive dispositions. Acknowledging and recognizing achievements is essential in sustaining effort and perseverance because, when students see where they are going and make visual progress, they are more likely to continue to stay in the game. These small reinforcements can mean the difference between engagement and apathy. They can give students a little boost that reinforces the idea that with effort and persistence they will succeed. Little certificates and notes such as those seen in Exhibits 3.1 and 3.2 go a long way to recognize, celebrate, and reinforce effort.

Exhibit 3.1 • Example of Goal Celebration Note

> **Congratulations!**
>
> Hong has learned doubles facts!
>
> He can:
> - Solve problems
> - Tell problems
> - Model problems
> - Explain the strategy

Exhibit 3.2 • Example of a Family Update Letter

> Dear Family,
>
> I wanted to let you know that Kelly is doing great with our chapter on dividing with remainders. She set a goal to master the skill, and she is doing a great job. Please see some of her attached work and notice how she has progressed over time. Bravo!
>
> Sincerely,
>
> Mrs. Johnson

Optimize Challenge and Support

In your Math Workshop Plus classroom, you will spend time thinking about ways to motivate students and offer the right level or challenge balanced with the right levels of support. Different students are motivated and challenged in different ways. The agency that comes with workstations is typically a motivator, so your focus may be on providing different ways to practice grade-level standards and ensuring access to appropriate scaffolds to meet the various needs of learners. Guided Math groups allow you to differentiate and meet students where they are, so the focus there may be more on motivation through exciting contexts, elevated levels of challenge, or nontraditional approaches. Make various tools and resources available so that students can get what they need to maximize every learning opportunity. Let's look at some ways to optimize challenge with support.

Tiering Math Problems

A great way to differentiate the degree of difficulty or complexity is to tier the practice problems (Tomlinson, 1999). Tomlinson talks about using the "Equalizer," a visual planning tool to help adjust tasks at "varied readiness levels." There are "nine continuums along which the difficulty level of lesson content, process or product may be located." Planning this way gives teachers an expanded "repertoire of ways they think about varying the challenge level of a specific task." Figures 3.4 through 3.11 show some of these ways. Tiered math activities are based on the same math concept or skill, but they have different levels of complexity. They can meet the needs and experiences of all students and allow everyone to work on grade-level concepts but with appropriate degrees of challenge.

For example, kindergarteners are expected to recognize their numbers within 20. A tiered activity by readiness, going from foundation to transformation, might be cards extended from 1 to 100 with students working within their range of proximal development (see Figures 3.4, 3.5, and 3.6).

Figure 3.4 • Kindergarten Tier 1 Activity, Numbers to 5

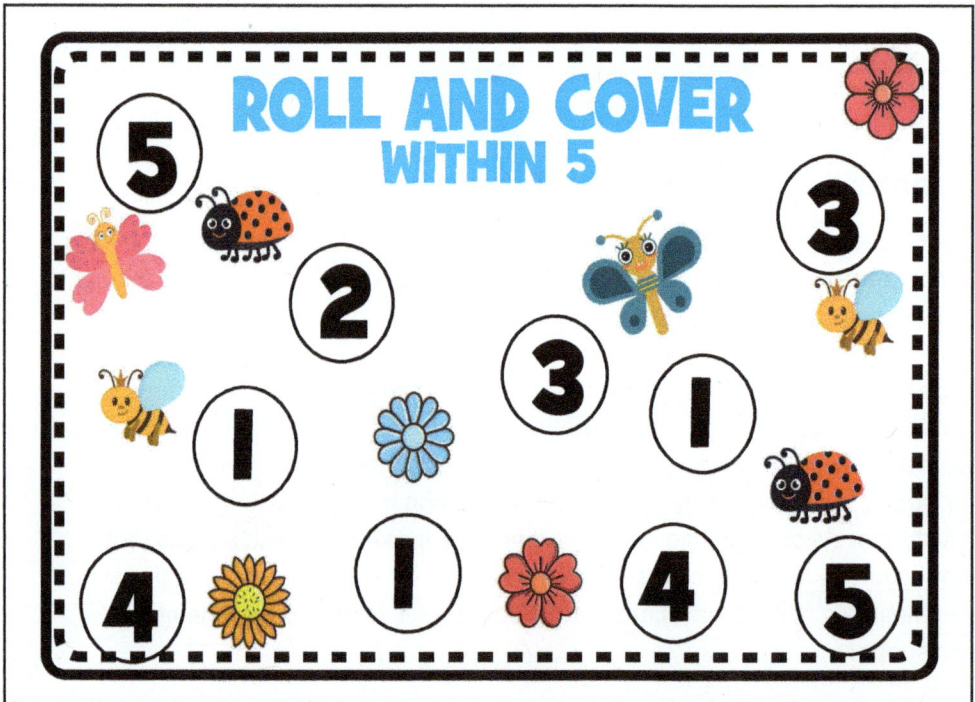

Figure 3.5 • **Kindergarten Tier 2 Activity, Numbers to 12**

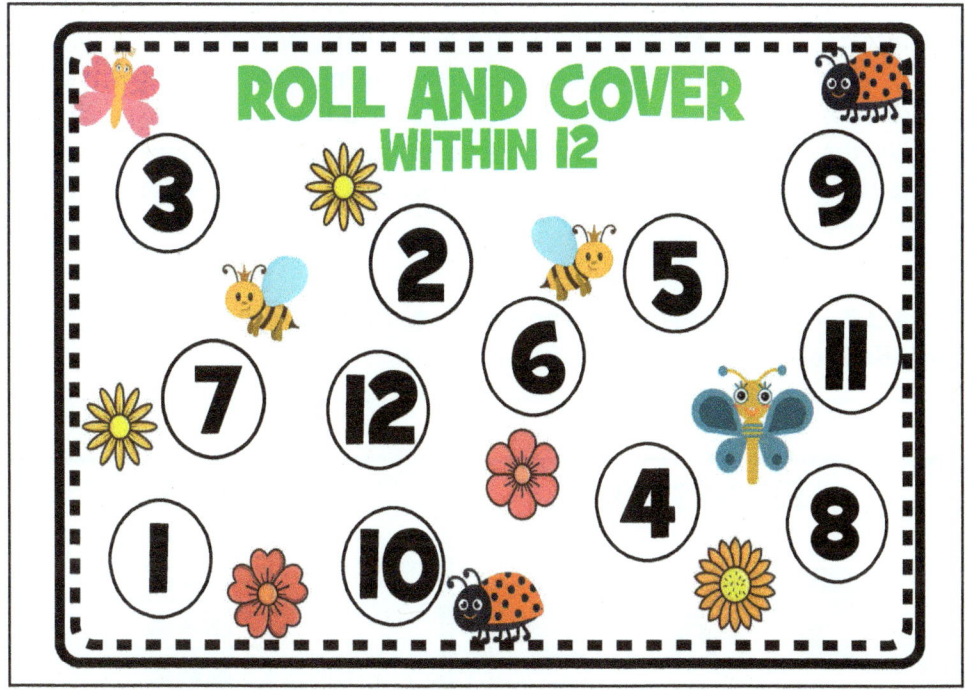

Figure 3.6 • **Kindergarten Tier 3 Activity, Numbers to 20**

In most states, third graders have to learn to multiply and divide within 100. Figure 3.7 shows an activity tiering that standard by going from a few facets to more facets. There are six cards in this game. Students can play with just two cards or up to all six. Students who are just learning multiplication might use the expression and match it with equal groups. You can add a layer of complexity by having the next set of students match the expression, equal groups, and the repeated addition statement. Some students are ready to find the matches to all the expressions.

Figure 3.7 • Tiering From Fewer Facets to More Facets—Grade 3 Multiplication

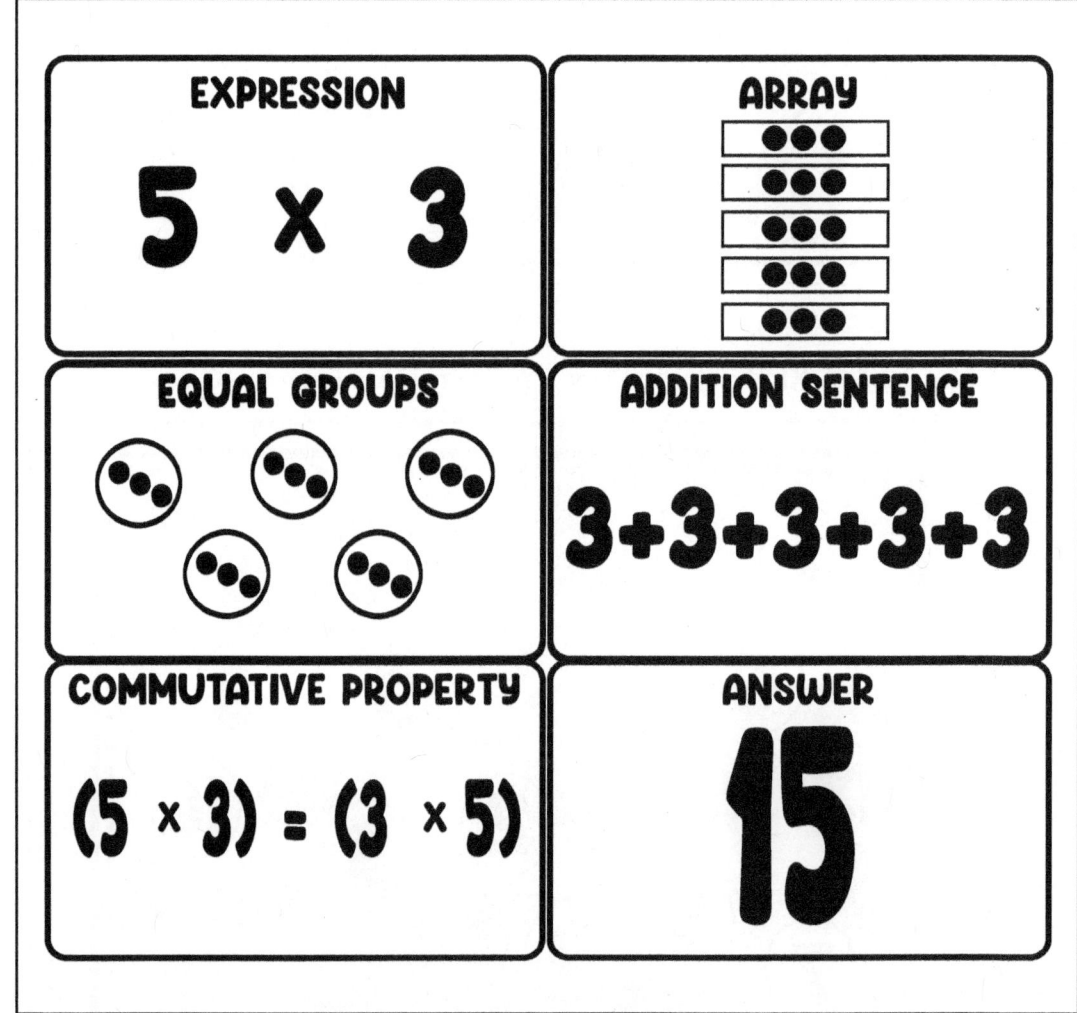

Figure 3.8 shows more concentration cards that can be tiered by facets, but also you can use the cards individually and have the students explore the concept from a concrete, pictorial, and abstract perspective. For example, some students might just play with the decimal card, where they pull a card and have to actually build the amount with coins. To extend this, they could multiply the number by 2 or 3 and build out that amount. Another set of students might pull a decimal card and have to shade in hundredths grids or ten frame paper to represent that decimal multiplied by a number that they spin or roll (in the beginning I would use numbers 1–4). This is a great strategy for students to see the break-apart strategy (partial products), where they multiply the tenths and then the hundredths and then add them together. Another set of students might be given blank cards, where they have to make a set of benchmark decimal cards from scratch.

Figure 3.8 • Tiering From Concrete to Abstract—Grade 6 Decimals and Percentages

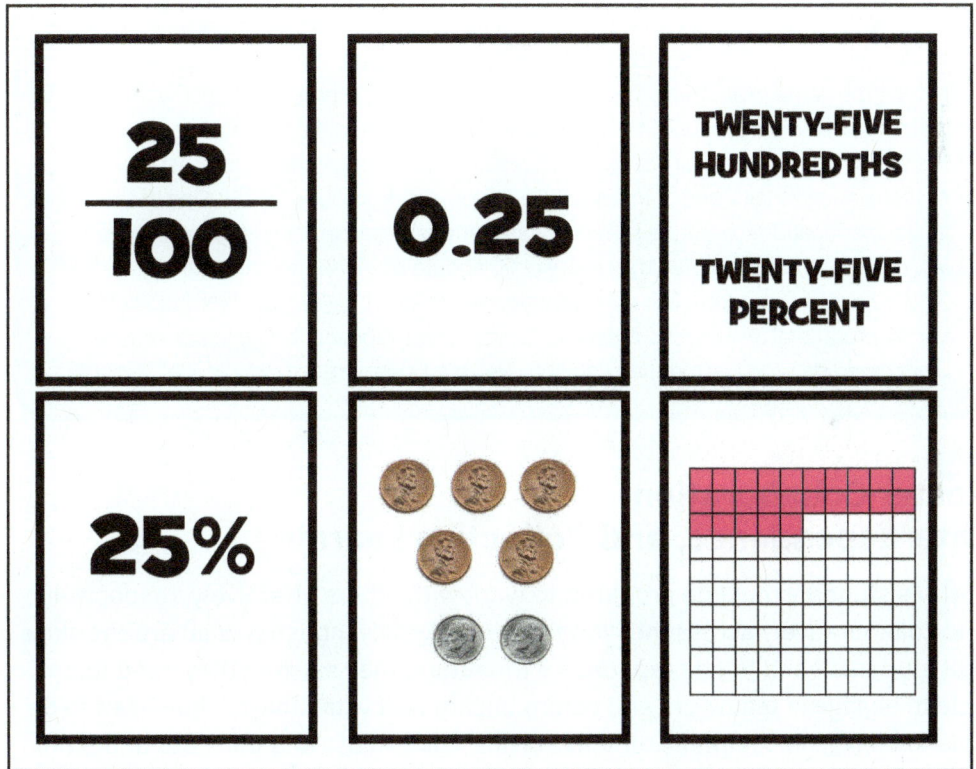

Connections to SEL

Self-Aware

Students should be self-aware of how they are progressing through the unit of study. They should be able to think about what resources and tools they need to do the work. Embedded rubrics give students the opportunity to manage their progress throughout the task—for example, putting a self-reflection emoji at the end of a task gives a student the opportunity to think about how they are doing with the work.

Equity Check

Optimizing Challenges

Looking through an equitability lens, it is important to vary demands and optimize challenges. Every child should get what they need. It is not about every child getting the same exact thing, but it is about every child getting challenged and learning how to persevere. This happens when teachers make the infrastructure of their lessons equitable. All the experiences need to be engaging, of interest to the student, and in their Zone of Proximal Development (Vygotsky, 1978). One of the goals of educational equity is that every student have an equitable opportunity to learn (IDRA, 2020). It notes that every child should be given the opportunity to learn "regardless of characteristics and identified need" and be given the "pedagogical, social, emotional and psychological supports" to be exposed to and supported to work on grade level so that they can achieve "high standards of excellence" (Goal 4).

Foster Collaboration, Interdependence, and Collective Learning

Today's students must be prepared to work with others effectively, respectfully, and collaboratively across the globe. When asked by industry, what are the skills that students need to be prepared for the future, they say that they need to be able to engage in teamwork and communicate well with others. They need to be able to engage in complex problem-solving, think in 3D, and think out of the box (Garskof, n.d.; Prepared Parents, n.d.). Students also need to have emotional intelligence, creativity, and cognitive flexibility (Prepared Parents, n.d.). They should be digital citizens, knowledge constructors, innovative designers, computational thinkers, and global collaborators (ISTE, n.d.)

Math Workshop Plus is premised on building a community of learners that learn and work together to become better mathematicians and good people that can get along with others, communicate, think creatively and flexibly, and innovate. From the beginning of the year, the focus is on building a strong community of students who are inquisitive, helpful, curious, and digital problem-solvers.

Set Community Goals in the First 20 Days

In a Math Workshop Plus classroom, you want to have a big kickoff every year that is about community building. Traditionally, a part of the first 20 days is devoted to helping students build a community of learners. We do lessons, reinforced with anchor charts, that teach students how to work in partners, work in groups, respect each other, listen to each other, and more. Drawing on the research about cooperative learning, we explicitly teach social skills, as we would any other skill, so that students understand what "get along," "share," and "be kind" actually mean.

Figure 3.9 is an example of a community agreement anchor chart to start the year. Students learn from the beginning that we value community. We talk in specifics rather than vague terms about "getting along." We lay out very clearly what it looks like, sounds like, and feels like. We refer to the chart heavily in the beginning and then throughout the year as needed to see if we are staying on track.

Figure 3.9 • Community Agreement Anchor Chart

Looks Like	Sounds Like	Feels Like
Working Together	"Please"	Great
Taking Turns	"Thank You"	Friendly
Being Friendly	"May I"	Happy
	"Could You Please"	Helpful

Getting Along Well Together (The Treatment Agreement)

Figure 3.10 is an example of what a poster like this looks like in a real classroom. In this case, the teacher chose to organize this as a T-chart. In Math Workshop Plus, students are expected to get along and work together to make great things happen. This is fostered by ongoing conversations about group work. Good groups don't just happen; they are planned for, cultivated, and maintained.

Figure 3.10 • A Classroom Poster of Great Groups

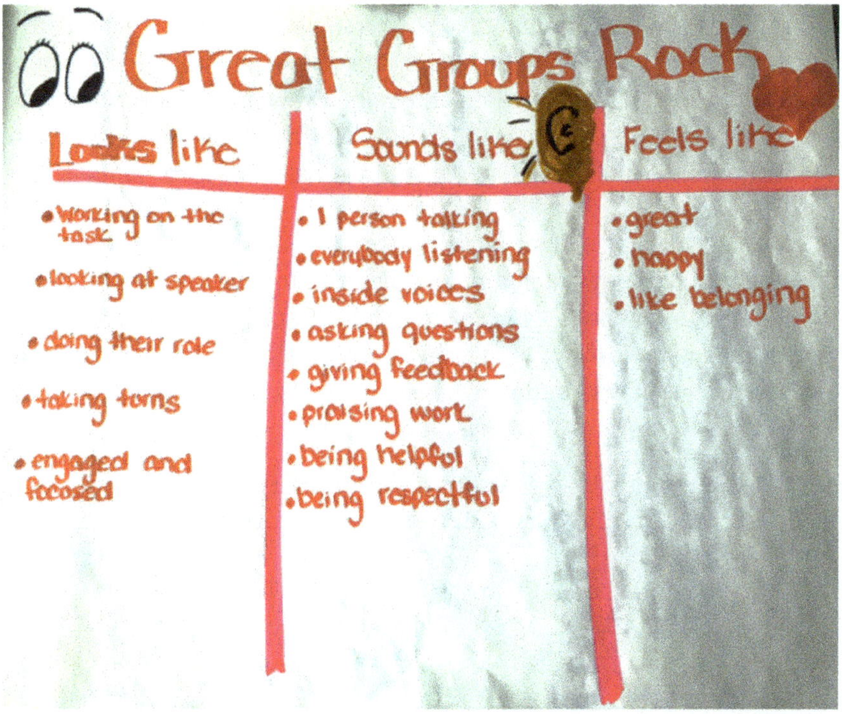

Use a Group Reflection Sheet

Group reflection sheets help students to reflect on how the group is working. One way to use them is to get the students to reflect once or twice a week on how they are working together. They should think about what is going well and how they keep up that momentum. Also, what needs to be changed? What do they need to work on as a group to be a better group? They discuss these as a group, and they turn them in to the teacher. You can see an example in Exhibit 3.3. We should have regular check-ins with groups and emergency meetings when called for.

Exhibit 3.3 • Example of Group Reflection Sheet

Group Reflection
Date: _____
Members: _____
Name: _____
Everybody worked: OK   Good   Great
Everybody shared ideas: OK   Good   Great
We took turns: OK   Good   Great
We listened to each other: OK   Good   Great
I would rate the work: OK   Good   Great
What went really well?
What would you do differently next time?

 Download this Group Reflection Sheet at https://companion.corwin.com/courses/MathWorkshopPlus.

More About Group Work

Collaboration and teamwork are skills that need to be explicitly taught. We draw heavily upon the work of Johnson and Johnson (1998), Kagan (1994), Cohen and Lotan (1995) and Cohen alone (1994) to inform what cooperative learning looks like in a Math Workshop Plus. Students can work with partners in ways that help them both grow. Cohen reminds us that the teacher must always plan for status treatments because status is always prevalent in a group (1994). A status treatment is intentional planning ahead of time so that everybody in the group has access to the task at hand.

Connections to SEL

Fostering Collaboration

In a Math Workshop Plus community, students are working together constantly. Math Workshop Plus is based on healthy relationships. Students work together, learn to communicate with each other, and support each other. We are working on navigating, building, and sustaining great relationships. We are also working on social awareness, fitting in, and being a stable, contributing, respected member of a flourishing community.

Equity Check

Fostering Belongingness

An equitable classroom ensures that there is a sense of belongingness, joy, happiness, risk-taking, and camaraderie. It is an environment where both the academic and social-emotional needs of all students are being met. As Madeline Hunter said, it is all about the "feeling tone" (1969). You know it when you see it, and anyone can pick up on it the second they enter your classroom. Equitable classrooms have a positive feeling and tone; they are inclusive and full of wonder.

Foster Belonging and Community

Creating a sense of belonging and community in your Math Workshop Plus classroom is essential for fostering an environment where all students feel valued, appreciated, and respected. If your students feel like they belong and that they have a safe space to work and learn, they are willing to take more academic risks. Building your learning environment with this goal creates a risk-taking environment so that students will try, fail, and try again without the fear of being ridiculed by their classmates. Cultivating a safe space where everybody feels like they can learn and help each other learn takes time, planning, and effort. Communities like this don't just happen but are cultivated through activities that allow students to get to know each other, trust each other, believe in, and root for each other. Classroom communities where everybody feels like they belong nurture students' emotional well-being and academic growth.

If your students feel like they belong and that they have a safe space to work and learn, they are willing to take more academic risks.

Building strong, caring classroom communities requires coming together regularly to check-in around norms. It means having routines and protocols for getting along and handling disagreements and engaging in community building activities throughout the year. When you invest in this, you save time in the long run. Your students will know how to incorporate new members of the community, even when they come in the middle of the year.

Your students will also learn to celebrate diversity in all of its forms and acknowledge and respect diverse backgrounds, abilities, experiences, languages, and ways of being and knowing. Look for ways to incorporate students' diverse backgrounds and perspectives into your lessons. When possible, your students should share their personal stories so that all voices are heard, appreciated, and respected.

Thriving communities have systems that work. If you are new to this model (and even if you're not), you may need to change your systems if they aren't working. When this happens, remember that students will need time to assimilate. This may involve resetting, modeling, and practicing. Students must know and understand the expectations of being a productive community member. They need consistency and support in working together, getting along, and thriving together.

Let's say, for example, your system for getting devices from the charging station was to release students all at once. You notice that this creates a bottleneck, and students aren't being kind or courteous. You reset and create a new system that sends students one group at a time or during a certain workstation choice or rotation. When sharing this change with students, you can reinforce the norms of respect, kindness, and cooperation in the classroom community. Remind them that, in a caring community, they are much more likely to feel safe and connected. They have to know how to solve problems because inevitably in a community problems arise. In this environment, students should know when it's appropriate to solve community problems on their own and when the teacher needs to be involved. If you are generous with positive reinforcement and take the time to share and practice conflict resolution strategies, your Math Workshop Plus blocks will run much more smoothly.

Ultimately, a classroom that fosters belonging and community gives students more than book knowledge. Knowing how to add fractions and multiply decimals is all a part of the equation, but in a Math Workshop Plus community, students learn the importance of relationships, how to get along with everybody, how to collaborate on important work, how to respect one another, and how to be a member of a community. This is knowledge that they will carry with them way beyond your classroom.

Connections to SEL

Fostering Relationship Skills

Fostering healthy relationship skills is a huge component of building a community. Kindergarten teachers are excellent at teaching students how to form positive relationships where they can communicate and get along well with each other. But, as you know, not all students master this before they come to you, so you will need to continue to support their development. By giving students opportunities to work with partners, teams, and small groups, they have authentic opportunities to master the art of listening, communicating, and compromise. They will also have organic opportunities to deal with conflict, and you can offer strategies to do this effectively. This leads to social awareness and builds an understanding of how their behavior impacts others.

Equity Check

Diversity, Equity, Inclusion, and Belonging

Diversity, equity, inclusion, and belonging (DEIB) are key concepts in building healthy, socially, emotionally, and academically safe, equitable schools and classrooms (Usanmaz, 2024). DEIB notes that diversity is about the differences between people; equity is about making sure everyone has fair and equitable access to the same opportunities and resources, inclusion means that everybody is a part of what is happening and they feel welcome and wanted; belonging is about feeling like a part of the group, a welcome member that is cherished for their individuality, accepted, and celebrated (Usanmaz, 2024). DEIB in the classroom is about creating a space where everybody can take risks, help each other, and learn together.

Offer Action-Oriented Feedback

Feedback is essential for learning. Feedback should help students feel encouraged about the process, motivated to continue, and guided to the next steps. When we were in school, feedback came in the form of red pen on our test papers. If the teacher was generous, she noted where we made our errors. Looking back, it's no wonder we just put our papers away and moved on—it was too late to change anything, and the grade was the grade!

As teachers today, we know better. You know that your feedback should be ongoing to promote perseverance. Your guidance will frame "learning as improvement rather than a fixed target" (CAST, 2024). Your feedback will help students develop self-monitoring skills and illuminate where the student is on *their* journey of learning, not where everybody else is. Use feedback to praise effort, encourage continued practice, and foster a sense of agency within your students. Use it to reinforce that learning is a journey not a destination and remind students that there are different milestones on that journey. As they have those "mini wins" along the way, they are getting better.

How do you most frequently offer feedback? Is it orally, such as through conferring, or written, such as notes in a journal or comments on various pencil-and-paper tasks and assessments? Considering the variability of your learners as well as the constraints on your time, perhaps using multimedia feedback could be more effective in some instances. Using multimedia can add to "engagement, interaction and most importantly accessibility and inclusiveness in the classroom dynamic" (Nguyen, 2021). If you've never tried using audio clips or visual feedback, you might love how it can capture and maintain students' attention and possibly save you significant time! It's as simple as recording your voice using any application on your computer or phone and sharing with students via your preferred platform. You can add a visual using your document camera if you want to show their work sample. You may be slow at first, but we've seen teachers who get so good at this that it ends up being faster and more effective, and the best part is it doesn't get lost like their papers! Think of the catalog you could have when it's time for conferences. It is a total game-changer. Just remember that, regardless of modality, effective feedback has specific characteristics.

Researchers have theorized several components of effective feedback (CAST, 2024; Wiggins, 2012), as summarized in Table 3.3.

Table 3.3 • Characteristics of Effective Feedback

Characteristics of Effective Feedback	Examples
Feedback must **reference the goal.** Reference both the goal and the learning intentions.	"This is a great answer. I like the way you are using the bar model, which is a model we have been working on in class."
Feedback must be **specific, constructive, tangible, and transparent.** It must be clear and easy to understand. Students should be able to know exactly what it is. Highlight on the feedback one or two things to think about.	"I want you to work on labeling your bar model so we know what the bars are representing."
Feedback must be **actionable and informative.** Feedback should answer the question "What do I do next?" Give some students a template or practice sheet for the next step.	"Next time, I want you to think about two things: 1. Do the bars represent the numbers and situation in the problem? 2. Is every part labeled?"
Feedback must be **kid-friendly and accessible.** Feedback should be written at the age level of the student. Feedback should also be spoken or transmitted through an audio device.	"If you had to rate your work: 1. You did everything correctly. 2. You did lots of things correctly but missed a few. Check on the list. 3. You must work on completing all parts of the problem. You only did one thing on the problem-solving list. What would you give yourself?"

Characteristics of Effective Feedback	Examples
Feedback must be **timely**. Feedback should be immediate so that students can use it right away.	The teacher gives a quiz and within 2 days gives the quiz back with feedback.
Feedback must be **frequent and ongoing**. Feedback should be continuous. Students should receive feedback daily. Whether it is through a quick exit slip, a mini conference, or a written note, the sooner they know, the quicker they can implement the feedback.	In a unit of study, it is important to plan for the feedback cycle. What does the assessment plan look like for the entire unit of study? Planning these things ahead means you are much more likely to do them.
Feedback must be **relevant and consistent**. It should make sense. It should be able to connect with an ongoing conversation. It should be accurate. It could have an icon that connects to the learning goal.	Feedback should connect to what the class and students have been working on. It should make sense. It should be accurate and directly connect to the goal.

Connections to SEL

Receiving Feedback

Students learn how to process the feedback you give them. This builds their awareness of where they are on their journey of mastering essential skills, and your feedback helps them think about and act on what they need to do next. Engaging in self-management involves goal setting and self-motivation. Students can think about how they feel about themselves as a mathematician, recognize their strengths, and also think about what challenges them. Over time, through your targeted feedback that encourages them, they can grow their self-confidence and their self-efficacy. Feedback helps students to be self-aware, and this self-awareness allows them to accept feedback not only from you but also from peers. Knowing that feedback is intended to be helpful not hurtful affirms that learning is a process that we all support each other through.

Equity Check

Equitable Feedback

Williams (2023) points out that "equitable feedback is not about giving [each student] the same quantity of feedback; it's about giving all [students] the same quality [of] feedback" (para. 5). Equitable feedback is about working with each individual student, meeting them where they are, helping them to improve with very specific information, guiding their next steps, and celebrating their successes. Just as you won't meet with every student the same number of times, you won't give everyone the same volume of feedback. Feedback is tailored to what you see from each student and is given as frequently as necessary to help them reach their goals.

Say Bye-Bye to Barriers When You . . .

- Create opportunities to stay motivated and determined
- Foster a goal-setting culture
- Expose all students to rigorous, challenging, grade-level work
- Allow students to "choose their own adventure"
- Create an environment of belonging, kind, curious learners
- Provide a culture of giving and receiving feedback

Action Plan

Assess your current success with Designing Options for Sustaining Effort & Persistence. Download the Chapter 3 Self-Assessment at https://companion.corwin.com/courses/MathWorkshopPlus.

Try It! Fostering Collaboration and Community

◇◇

In what ways are you fostering collaboration and community in your classroom? What are some new ideas you have after reading this chapter? Try one of them.

 Connections

◇◇

Post an artifact of something that you have done to sustain effort and persistence on social media. #mathworkshoPLUS

Chapter 4

Design Options for Emotional Capacity

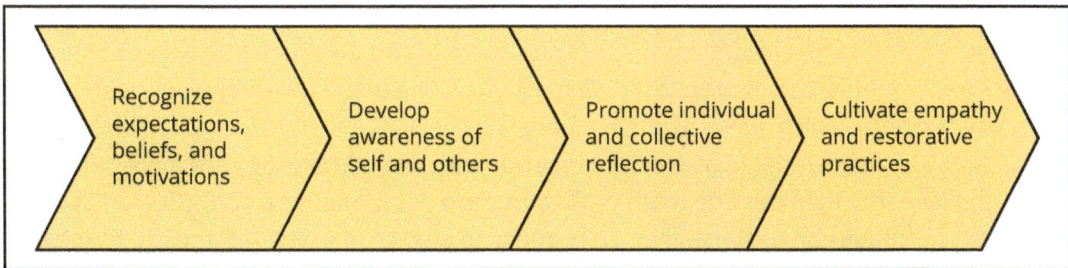

Math has a bad reputation. The word itself can evoke fear, anxiety, and stress. Many students, parents, and even teachers, associate negative feelings with math. In 2008, Jo Boaler noted that twice as many people reported hating math than all other subjects combined! As teachers, we are not immune to this. Many of us had negative experiences in our own education that left us feeling uneasy about math. As students move through the grades, they may have experiences that make them feel confused and unsure of themselves. They may notice that others can produce answers faster than they can. They may complete problems and procedures that they don't understand, and they may become reluctant to participate for fear of being wrong and feeling embarrassed. You may feel powerless to change the minds of students, but you *can*, and it starts with understanding the role that math disposition plays in how students learn math.

Do you know how your students feel about math? We believe that knowing this is crucial, as these feelings will affect how students participate, their willingness to persevere, and how comfortable they will be working with others and taking risks (Farrington, 2013). Throughout this chapter, we'll consider practical ways to *Design Options for Emotional Capacity*, and we'll highlight key connections to the SEL competencies. The following questions will help assess your starting point in each facet of this dimension and help you leverage Math Workshop Plus to remove barriers.

> 1. *Recognize expectations, beliefs, and motivation*: How do students feel about math, and how do you know?
> 2. *Develop awareness of self and others*: What supports are in place to develop and encourage healthy emotional responses?
> 3. *Promote individual and collective reflection*: Do you have systems that guide students to monitor progress?
> 4. *Cultivate empathy and restorative practice*: How can you facilitate routines and a debrief to build, maintain, and restore community?

Regardless of the grade you teach, how students perceive their ability to succeed in your math class matters. Math disposition is one of the five strands of math proficiency, which work together synergistically (National Research Council [NRC], 2001). When students possess a negative math disposition, they are more likely to believe that they can't do math or that they are not a "math person" (Boaler, 2015). This attitude can diminish their willingness to engage in problems and often results in students giving up easily, shutting down, or waiting to be rescued by an adult. If this sounds familiar, we are not surprised. The good news is there are simple things you can improve dispositions, cultivate confidence, and encourage risk-taking.

Even if you do feel like you have a strong sense of where students are in terms of their math disposition, we recommend collecting disposition data. There are many free instruments that are easy to administer and offer valuable insights into student mindsets. The data can point you to students who may need more targeted efforts to improve their beliefs about themselves.

Recognize Expectations, Beliefs, and Motivations

Positive math disposition is often linked to Carol Dweck's (2007) work on growth mindset, which has been widely accepted and applied in the world of education. Dweck's work suggests a correlation between our beliefs about ourselves and the outcomes we experience. She theorizes that if we experience failure and believe that no matter what we do, we will continue to fail—we possess a "fixed mindset." In contrast, if we experience failure but believe that with practice and effort we will improve, we possess a "growth mindset." In her book *Mindset: The New Psychology of Success*, Dweck states,

> The passion for stretching yourself and sticking to it, even (or especially) when it's not going well, is the hallmark of the growth mindset. This is the mindset that allows people to thrive during some of the most challenging times in their lives. (2007, p. 10)

This has resonated in the math education community, where struggle and challenge are common and beliefs about being a "math person" and *not* being a "math person" are pervasive and damaging. In 2015, Stanford University professor Jo Boaler connected Dweck's work to mathematics in her book *Mathematical Mindsets* (2015), which highlights strategies to transform math dispositions and cultivate growth mindsets. She details how to create a culture where all students believe they are capable of succeeding in math, shifting from "I can't do this" to "I can't do this *yet*."

Boaler has evangelized the power of mistakes in math and cites findings from Moser et al. (2011) showing how the brain grows when you make mistakes. In Math Workshop Plus, how *we* view mistakes, how we react and handle them, and how we help students frame them can make or break their impact on motivation. In her 2016 TEDx Talk, Boaler discussed the changes needed to cultivate growth mindsets and positive math dispositions. In a video from her summer camp for middle schoolers, students are shown singing their version of Taylor Swift's "Shake It Off" in relation to mistakes. We love sharing that video along with the poster in Figure 4.1.

Figure 4.1 • Mistakes Poster

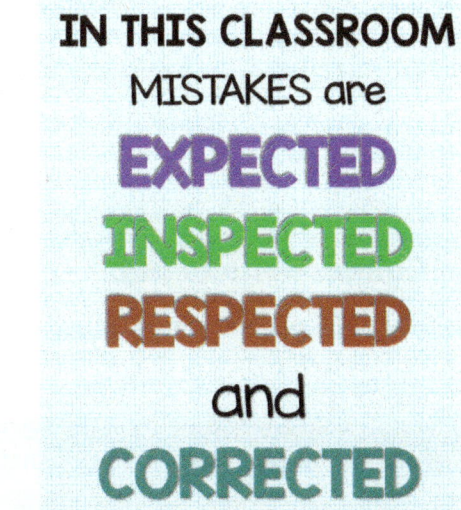

You may notice that Boaler's suggestions align well with UDL, promote equity, and are consistent with the strategies we are suggesting. As you continue to explore this book, watch for connections to evidence-based shifts. Boaler (2016) makes these suggestions:

- Use open math problems (those that have more than one solution) and allow students to share different methods and ideas:
  - The area of a rectangle is 24 square units. What could the dimensions be?
  - The sum of three numbers equals 17. None of the numbers is greater than 9. What could the numbers be?
- Make math visual
- Teach with mindset messages and normalize mistakes as part of learning
- Encourage collaboration, sharing thinking and manipulative use
- Deemphasize speed and focus on depth and creativity

This book has many ideas to support your efforts to enact Boaler's suggestions, but one is a favorite. Several years ago, when teaching her graduate course on Math Workshop, in addition to the "Shake It Off" video, Alison had her students watch a *Sesame Street* video about "The Power of Yet." As Janelle Monet sang about not being able to do things *yet*, everyone saw the power of this message.

A principal who was taking the course loved the concept and ran with the idea as a schoolwide mantra. He built it into his assemblies, staff meetings, and parent interactions. He hung posters in the halls and often added "yet" to the end of not just his own sentences but the sentences of others. Soon, the Power of Yet was everywhere, and it became common to hear students and teachers adding "yet" to sentences that typically started with "I can't" or "I don't know how to" or "I haven't." While it may have three letters, the difference in the mindset of those adding "yet" was so powerful that we referred to it as our "secret sauce." We hope it can be your secret sauce too!

Connections to SEL

Fostering Self-Management & Self-Awareness

Math classrooms can inherently challenge students in the social and emotional realm. Struggle and confusion are often triggers that put pressure on fragile SEL competencies, particularly self-management and self-awareness. This may appear as shutting down, acting out, losing focus, lacking motivation, showing visible signs of stress, and engaging in negative self-talk. We have the power to normalize struggle as part of learning and to teach math with a focus on student thinking and understanding rather than just answers and one way of solving.

Try It!

Examine Your Mathematical Mindset

Before we discuss strategies you can implement to improve student mindsets, take a moment to examine *your* mathematical mindset by answering the following questions:

1. How do I feel about math?
2. Do I enjoy teaching math, or do I avoid it? Why?
3. How confident am I in the content and standards I teach?
4. Do I believe that all students are capable of learning grade-level math?
5. What messages do I send about math, about whose ideas matter, and about what it means to be good at math?

If your answers to the Try It! questions indicate that your mindset needs some adjustment, you aren't alone. Many of us have personally struggled with math as learners, and those feelings often stick around well into adulthood (Boaler, 2015). While this is nothing to be ashamed of, it is important to recognize these feelings so you don't inadvertently pass them along to students. This phenomenon and its implications on student learning have long been studied. In a 1978 article, Mihalko stated that

> [Mathematics teachers] cannot be expected to generate enthusiasm and excitement for a subject for which they have fear and anxiety. If the cycle of mathophobia is to be broken, it must be broken in the teacher education institution. (p. 36)

You may be determined to provide students with a positive math experience regardless of what your own experience has been. Table 4.1 offers a brief exercise that can help to reveal ways your own disposition can color your decisions and actions, whether you are aware of it or not. This involves unpacking some common teacher moves that may inadvertently undermine student confidence and damage math dispositions. Note that these examples do not reflect bad intentions. We highlight them because they make it more difficult to promote expectations and beliefs that optimize motivation. The recommendations we share suggest simple shifts to move from negative to positive.

Table 4.1 • Teacher Moves That May Limit Beliefs and Motivation

Teacher Move	Possible Inadvertent Message(s)	Recommendations	Notes
Asking if student would like to "phone a friend" for help answering a question	• I don't think you can do this. • There is no part of this that you understand. • You're not fast enough.	• Ask student what tools or resources they can use. • Ask students what parts they do know and/or use questioning to reveal what they know. • Tell student you will give them think time and come back to them.	
Picking the first students who raise their hands	• To be good at math, you need to answer quickly.	• Use turn and talk before sharing. • Allow quiet think time for everyone before seeking answers. • Use sticks to select students.* (*If you use this method, we strongly recommend using turn and talk prior to calling on students so that they can have more confidence in their answer before sharing.)	

Teacher Move	Possible Inadvertent Message(s)	Recommendations	Notes
Just accepting answers and moving on	• "I don't care about your thinking or how you solved the problem." • There is only one way to solve. • Math is about answer getting. • "You missed your chance to learn this, and we are moving on." • Math is about grades and correct answers, not about learning and understanding. • "I don't expect *you* to be able to get this," or "It's not important for *you*."	• Always ask students how they solved. • Allow multiple students to share their strategies. • Highlight different strategies. • Ask any of the following: Why? Are you sure? How do you know?	
Not expecting or allowing students to correct mistakes	• "You missed your chance to learn this, and we are moving on." • Math is about grades and correct answers, not about learning and understanding. • "I don't expect *you* to be able to get this," or "It's not important for *you*."	• Normalize the practice or error analysis and reflection. • Highlight the value of understanding what is right about our work and where we went wrong. • Work with student in a small group to provide intervention before trying again or making corrections.	

(Continued)

(Continued)

Teacher Move	Possible Inadvertent Message(s)	Recommendations	Notes
Focusing on what the student does *not* know rather than what they *do* know	• I am bad at math. • I get everything wrong.	• Focus on, and illuminate, what the student does know. • Ask the student where they think they broke down and help them to see all that they did correctly.	
Working with students to complete work that is intended to be completed independently	• I don't believe that you can do this on your own. • The other students are smart, but you aren't so you need my help.	• Promote independence by encouraging all students to use their math tools, anchor charts, and peers for collaboration. • Normalize struggle and model what to do when you get stuck.	

Did anything hit home for you? If so, we've been there. Because our actions are often intended to support students and protect them, we rarely realize the messages that we may send. Unfortunately, impact trumps intent. In addition to the recommendations in Table 4.1, this chapter will offer examples and strategies to support your efforts to optimize motivation. When we know better, we do better, so let's get to it!

Equity Check

Supporting Independence

Students who receive support services are less likely to have the same opportunities as their peers to work independently. Not only can this foster negative dispositions, but in a Math Workshop Plus environment, it can translate to missing out on playing games, engaging in productive struggle, and having the chance to make choices for themselves. If this is happening, think about ways to support independence, especially if you see signs of learned helplessness. Every student deserves to experience all aspects of the workshop and workstations, and it's up to us to ensure equitable experiences for students, even if we're nervous about releasing control or allowing students to struggle.

Develop Awareness of Self and Others

Students thrive when given some control and can move around as needed and match their attention spans with tasks. With UDL, students have even more access to what they need and more agency to select it. These elements decrease the odds of frustration, curb off-task behaviors, and foster more success both in academics and in SEL competencies. Even with UDL in place, taking time to intentionally develop and manage healthy emotional responses and interactions pays off.

Reflect on how your current learning environment is designed. Consider the following questions:

- Are the norms and expectations posted?
- What happens when students aren't meeting them?
- Do the norms promote and support coping skills?
- Do students know what to do if they get stuck?
- Do they have strategies to use if they become frustrated, confused, or upset?
- What systems support positive social interactions?

Note where you need to focus and shift your attention to enacting proactive solutions. What needs to be in place for students to meet expectations? For example, have you made anchor charts with students that outline norms and expectations? Have you practiced how workshop should look and sound, including how to transition, how to be a good partner, and how to appropriately react? Have you discussed what to do when stuck? What strategies have students learned to solve problems *before* frustration sets in?

Take time to review the following samples in Figures 4.2 through 4.5. We've selected areas where we see a need for guidance and reminders and where challenges may test SEL competencies. Which ideas would make a difference in your classroom? How can you involve students in your next steps?

Figure 4.2 • What to Do if You Feel Stuck

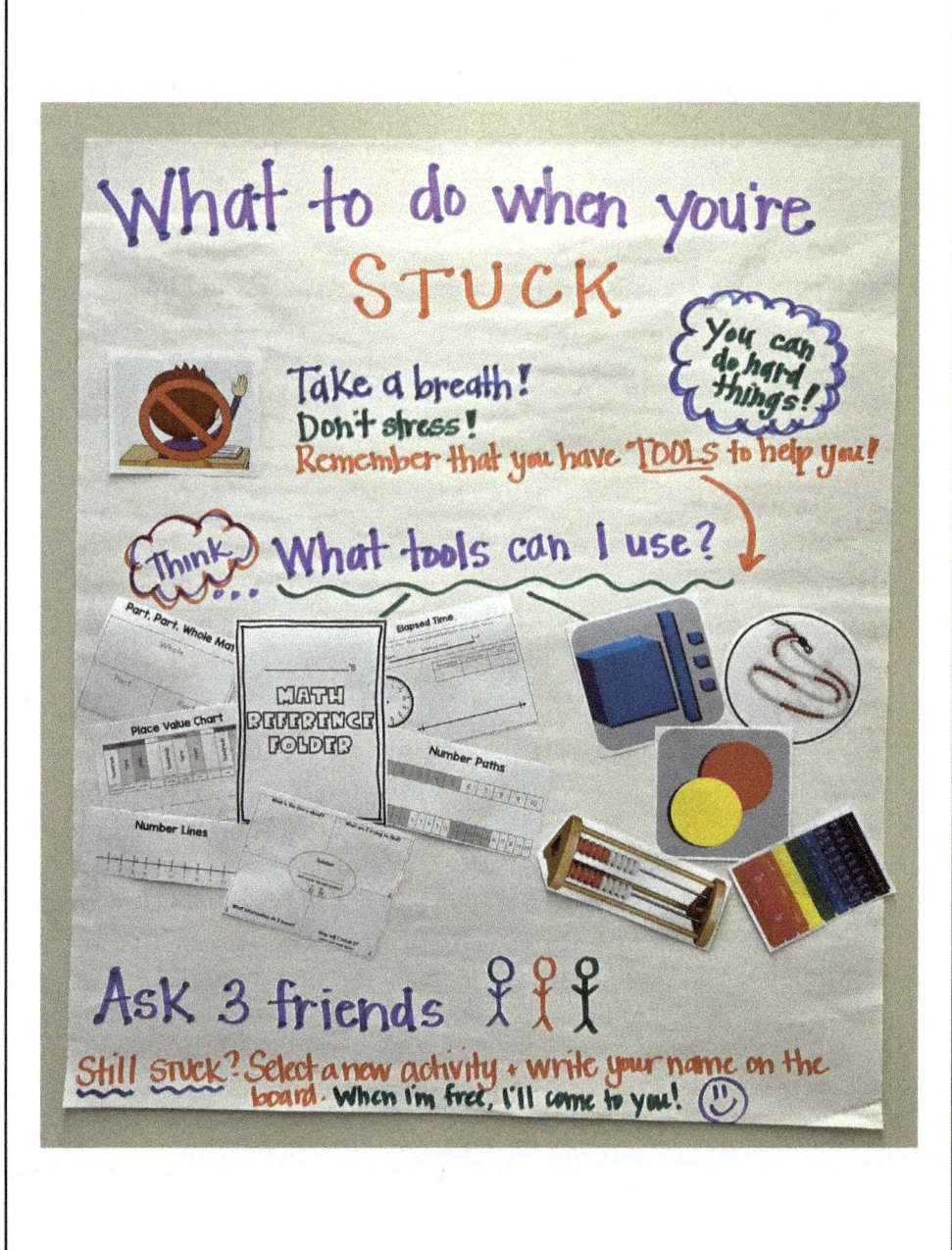

Figure 4.3 • What Do Good Mathematicians Do When They Get Stuck?

[Poster: "What do good mathematicians DO WHEN THEY GET STUCK?" with nine panels — LOOK AT IT AGAIN; ASK for HELP; LOOK at the ANCHOR CHART; DRAW A MODEL; Use easier NUMBER first. 1 5 3; THEY TRY it a different WAY; KEEP Trying; Double check the model and the MATH; NEVER GIVE UP.]

Source: Newton, N. (2022). Math Workshop Booklet. Newton Education Solutions.

 Download this Poster at https://companion.corwin.com/courses/MathWorkshopPlus

Figure 4.4 • Anchor Chart Outlining "What to Say to Yourself Instead of I Don't Know or I Can't"

Figure 4.5 • Anchor Chart Outlining Things to Remember When Playing Games

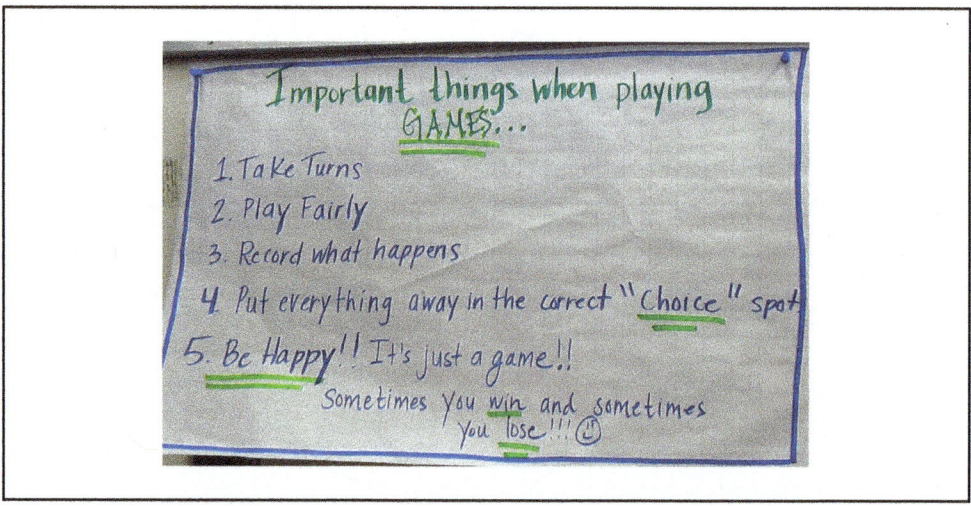

Behavior is communication, and in a math classroom, understanding the root cause is the key to adjusting your systems. If students feel unsuccessful, they may form incorrect assumptions about their ability. Don't forget that behavior issues can also stem from boredom, so ensuring that all students are challenged is important.

Equity Check

Unintended Consequences of Behaviors

If you have experienced challenges with behavior, regardless of whether you run a workshop or not, take a look at who frequently needs redirection. Who misses the opportunity to partake in games, work with peers, or make choices due to their behavior? Do you see trends by gender, race, neurodiversity, or ethnicity? Research consistently shows that there are "well-documented racial, gender, and income disparities in disciplinary outcomes" (Gregory et al., 2010, Losen et al., 2015, and Skiba, 2015, as cited in Welsh & Little, 2018, p. 752). Further, "the disparities in disciplinary outcomes are fairly consistent across all settings and grades, indicating a systemic problem that starts as early as preschool" (Welsh & Little, 2018, p. 752). Take some time to consider possible root causes of behaviors and brainstorm how the classroom environment better could support students.

Develop and Manage Healthy Emotional Responses and Interactions

When you use games, students experience engaging practice that is designed to be fun, low pressure, and enjoyable. For students who think games are about winning and losing, this isn't the case, and losing can be very unpleasant. We love teachable moments, and when we introduce games, we tend to have lots of them!

Games create ideal conditions for students to develop and manage healthy emotional responses and interactions. There are different types of games, such as competitive, collaborative, adaptive, and different modalities, such as digital, tactile, and paper-based. Students can participate in games independently, with partners, or groups, and games can be adapted to add or subtract elements to create differentiated experiences. Gameplay generates feelings and emotions that engage several of the SEL competencies which intersect with this dimension of UDL. Taking advantage of this synergy is like getting a two-for-one deal!

CAST suggests we universally design to provide differentiated models, scaffolds, and feedback for the following:

- Managing frustration
- Seeking external emotional support
- Developing internal controls and coping skills
- Appropriately handling and shifting self-judgements such as "I'm not good at math"

When looking at the SEL competencies, we see overlap. CASEL suggests offering opportunities for students to develop and strengthen the following:

- Self-Awareness
  - ✓ Identifying one's emotions
  - ✓ Linking feelings, values, and thoughts
  - ✓ Experiencing self-efficacy
  - ✓ Having a growth mindset
- Self-Management
  - ✓ Managing one's emotions
  - ✓ Identifying and using stress-management strategies
  - ✓ Exhibiting self-discipline and self-motivation
- Relationship Skills
  - ✓ Communicating effectively
  - ✓ Seeking or offering support and help when needed

What specific actions can you take to support these dynamic processes? Once again, we go back to the First 20 Days and the importance of establishing norms, engaging in situational modeling, and role-playing using examples and nonexamples. These experiences, and the points you want students to take from them, should be documented on anchor charts students can refer to. Some of the most helpful charts we've created to support healthy emotional responses and interactions relate specifically to game play. Refer to Figure 4.6 and Figure 4.7 for examples.

Figure 4.6 • Anchor Chart Outlining "How to Start a Game"

How To Start a Game

Dice Games

Each player takes a turn to roll the die. The player with the highest roll goes first. If there are more than 2 players, the second highest roll goes next. The lowest roll goes last.

Card Games

Each player takes a turn to pick a card. The player with the highest card goes first. If there are more than 2 players, the second highest card goes next. The lowest card goes last.

Money Games

Each player takes a penny and tosses it into the air and catches it. The coin is then flipped into that player's hand. If the coin lands on HEADS, that player goes first. If more than one player lands on HEADS, then flip again.

Spinner Games

Each player takes a turn to spin the spinner. The player with the highest spin goes first. If there are more than 2 players, the second highest spin goes next. The lowest spin goes last.

Figure 4.7 • Anchor Chart Outlining "How To Play a Game"

How to Play a Game

1. Get all the materials.
2. Find out who plays first, next, etc.
3. Get started right away!
4. Take turns and play fairly.
5. Be a good sport.
   Remember:
   You win some and you lose some, but the fun is in playing the game!
6. Clean up materials.
7. Put things back where they belong.

Connections to SEL

Personal SEL Goals

Having students set personal SEL goals is a great way to help students tune in to this part of their development. Students can use and/or create checklists based on the anchor charts in Figures 4.6 and 4.7 and track their progress. If appropriate, you may use peer evaluations if the goal is related to being a good partner, communicating effectively, or how to win and lose.

Promote Individual and Collective Reflection

In a high-functioning Math Workshop Plus classroom, we offer daily opportunities for students to assess themselves and reflect, individually and together. These assessments and reflections may be about the math they learned that day, the way they interacted with peers, how well they followed classroom norms, how effectively they managed their emotions while playing a game, how independently they worked, how ready they feel for assessment, or any number of other things. This provides us with valuable insight into how students perceive themselves, how they are feeling, and what they may need from us.

Opportunities for self-assessment and reflection pop up throughout the workshop block. Let's go through each component and look at examples you can easily integrate. If you are new to this, pick one place to start so you don't get overwhelmed.

1. The Launch

Students feel confident to participate in the launch thanks to the low floor–high ceiling tasks. For this component of the workshop, students may reflect on the following:

- Which strategy shared was most efficient and effective?
- How might I approach a similar problem differently?
- On a scale of 1 to 5, how well did I: listen, participate, explain my thinking, and so on?
- What did I enjoy about the routine?
- How was my strategy different from someone else's? How was it the same?

Depending on how you hope to use this information, you could have students reflect and self-assess on individual whiteboards, journals, on an exit ticket, or even on a sticky note. Since we're stretching ourselves to think through the lens of UDL and equity, how else could this look? Each option in the previous bullet list involves students responding in writing, but this may not be the most effective modality for all students. Offering a choice here would help ensure that all students can fully respond. How might you incorporate the following options?

Students choose to

- Record their reflection on a journaling app
- Circle a rating on a ready-made sheet
- Talk with a partner to share their reflection
- Select an image that matches their response (emoji, etc.)

2. The Mini Lesson

Five minutes into the math block, and we've already had so many great opportunities for students to reflect and self-assess! How can you keep this going during the mini lesson? Some ideas include these:

- Thumbs (up, down, sideways)
- Cups (red, yellow, green)
- Fingers (1, 2, 3, 4)
- Student response systems (clickers or other digital responders)
- Exit tickets

Connections to SEL

Reflection Shouldn't Be a Stressor

Watch for signs that students may be uncomfortable with publicly sharing their reflections. If this causes stress, students who struggle with self-awareness or self-management may shut down or be dishonest in their assessment for fear of embarrassment (Bledsoe & Baskin, 2014). Be sure to offer alternatives to all students so everyone can fully and authentically participate.

3. Workstations and Guided Math Groups

Following the mini lesson, students may either move into workstations or join you for a small, Guided Math group. If you use a printed menu, include a self-assessment scale such as the one in Table 4.2. This prompts students to reflect after each workshop and rate themselves. Remember that the reflection can vary to focus on norms, math, goals, or anything that is relevant that day.

Table 4.2 • Sample Math Workshop Self-Assessment

5	4	3	2	1
I was fully focused the whole time. I gave my best efforts.	I was fully focused most of the time. I gave great effort.	I got a bit off-task but regained focus by myself.	The teacher had to remind me to get back on task.	I didn't do my best work. I was off-task often.

4. The Debrief

By design, the debrief prompts reflection and self-assessment. As the workshop ends, we ask students to share what they learned. If you have trouble fitting the debrief in, elect a timekeeper to help you reserve two to three minutes. This communicates that students are accountable for their learning and that you're interested in hearing their thoughts, perceptions, and feelings about their experiences. There are many ways to do this to account for variability and keep it fresh. Table 4.3 offers a few ideas to maximize reflection during the share, without taking up more than a couple of minutes.

Table 4.3 • Quick and Easy Strategies to Promote Reflection During the Share

Strategy	SEL Benefit	Description
Turn & Learn	Building social awareness and relationship skills	Students share a "jump" (triumph), "bump" (challenge), or "something they learned" (ah-ha) with a partner. If called on, students share the reflection of their **partner** instead of themselves.
Exit Ticket With Rating Scale	Building self-awareness and self-management	Students complete an exit ticket that includes a rating scale about how confident and successful they feel (may be pre-made, written on a sticky, or in digital form). Students may indicate that they would like additional time with the teacher, additional practice, or an additional challenge.
Journal Entry	Building self-awareness and self-management	Students reflect in their personal journals (paper or digital—option for speech-to-text, visuals, etc.). Some prompts we like: "I used to think, but now I know _____." "Today I feel _____; tomorrow I hope to _____." "Today I helped myself by _____." "I hope that tomorrow I can _____."

Increase Awareness Around Progress Toward Goals and How to Learn From Mistakes

As students assess themselves and reflect on their progress, we see shifts in mindsets about mistakes. Focusing on progress over perfection and normalizing mistakes as part of learning allows students to see themselves differently. This doesn't happen at the same rate for all learners, so we must reinforce, model, and remind students that FAIL stands for First Attempt In Learning.

The following routines shine a spotlight on mistakes, what's right about them, and what we can learn from them. One is called My Favorite No, and the other is Fix It Friday. Both involve identifying and correcting errors, but they also show that mistakes are often a small piece of otherwise excellent work. We've shared brief overviews in Table 4.4, but you can find videos with a quick internet search.

Table 4.4 • Routines That Normalize Mistakes

	My Favorite No	**Fix It Friday**
What	Warm-up routine to showcase common errors while highlighting what is correct	A routine where students correct errors from their work that week
Why	✓ Normalizes mistakes ✓ Highlights what's *right*, even if the answer is wrong ✓ Strengthens procedural fluency	✓ Normalizes mistakes ✓ Highlights what is *right*, even if the answer is wrong ✓ Builds an understanding that learning happens on a continuum, and as we learn more, we can make corrections
Materials	Choice of index cards, half sheets of paper, or shared slide deck if using devices	Students select a sample of their work from the week to correct Quiz Homework Exit Ticket Problem of the Day/Week

	My Favorite No	**Fix It Friday**
How	1. Present a problem for students to solve (i.e., an equation). 2. Students solve using any method they choose. 3. Students submit their finished problem. 4. You sort responses looking for common error patterns. 5. Select a work sample to share on the screen. ***IMPORTANT: Recreate student work to avoid identifying whose work is displayed.** 6. Ask students to notice what's correct about the work. Select several students to share observations. 7. Ask students where they think the error is and facilitate a discussion highlighting why the error is common and how to avoid it.	1. Students select a work sample. 2. Students redo select problems either on a separate sheet, device, or form you create. 3. Students show/explain their original error and what they did differently this time. 4. Students share what they did during the week to better understand the concept so they could identify the error (could be Guided Math group, computer-based practice, manipulatives, self-checking workstation, etc.). 5. Students rate how they feel about the concept now.
UDL Considerations	• If students cannot see the screen, use a shared slide deck for access on personal device • If students cannot share their thoughts orally, allow them to submit them using technology or on a whiteboard • Allow turn and talk/group discussion before sharing to support language • Provide sentence frames.	• Offer alternative options for students to complete Steps 3 and 4. For example, submit an audio or video recording of their explanation versus writing. • Offer language frames to support explanations. • Allow students to show their corrections visually or using manipulatives. • Offer rating scales that can be circled or checked off to decrease writing/language load.

(Continued)

(Continued)

	My Favorite No	**Fix It Friday**
SEL Considerations	• Be strategic when selecting work to avoid triggering students who are fearful of being wrong. • Leverage this as an opportunity to strengthen self-awareness and reinforce growth mindset messages. • Reinforce self-management by spotlighting students who used tools or organized their work particularly well. • Build social awareness by selecting students to share who may need practice focusing on the strengths of their peers.	• Initiate this routine by having students find the teacher's mistakes and correcting them to model self-awareness (growth mindset) and self-management (mistakes are nothing to get upset over). • Highlight the value and feeling of relationship skills (seeking support and help when needed) by allowing students to share their corrections with peers for feedback before submitting.
Keys to Success	• Focus on what is *correct* and stress that the error is *common* so students don't feel like they are the only ones making the mistake. • When possible, share an error you anticipated but did *not* see. • Use this routine to build confidence and clarify misconceptions in a safe way.	• Avoid overwhelmingness for certain students by assisting them in selecting a work sample. • Leverage your Guided Math groups and workstations to ensure students are prepared to make corrections (re-teach, practice, etc.). • If you are in a grade-centric environment, consider allowing students to regain points after making corrections. • Use this routine to reinforce the goal of proficiency, not grades.

Cultivate Empathy and Restorative Practices

While math class may seem like an unlikely place to cultivate empathy and restorative practices, in a Math Workshop Plus classroom there are countless opportunities. As seen in Chapter 3, your students will collectively develop an agreement regarding classroom norms and expectations. This "Treatment Agreement" serves as a visual reminder for students and as a guide you can refer to when challenges arise.

If you teach younger students, challenges may look like simple squabbles about which partner to work with or frustration over losing a game. If you teach older students, challenges may involve more serious issues such as the math trauma that students carry or the impact of years of confusion on the way that students view themselves as knowers and doers of math. If you ever struggled in math, this may evoke feelings of empathy from you. How do you cultivate conditions where students develop and express that type empathy for each other?

Learn From Others' Perspectives and Repair Harm

Think about the debrief at the end of Math Workshop Plus. If issues have arisen during class, such as those mentioned, the debrief offers the perfect space to have students share their feelings, express their perspectives, and discuss specific exchanges that do not align with the Treatment Agreement. As some students share, the other students learn to listen and consider the perspectives of those who are experiencing math in a different way than they are.

For example, if two students were playing a game and one got frustrated with the other because they work more slowly, the slower student may express that they felt embarrassed or upset. If this is done in a circle, or even in a morning meeting the next day, other students can discuss times when they've experienced similar feelings. The restorative circle is used to bring students back to the importance of the treatment agreement and help students to renew their commitment to maintaining a kind and supportive learning environment where differences are respected.

Since Math Workshop Plus is predicated on inclusivity and honoring all learners, this component can and should play a significant role in your learning community. Take time during the First 20 Days to present mock scenarios and role-play how students feel and why addressing their feelings matters. Normalizing restorative practices in itself builds empathy and contributes to the development of the whole child.

Say Bye-Bye to Barriers When You Know . . .

- Mindset matters for students *and* teachers
- Your actions can support or undermine positive math dispositions
- Math is not about speed
- The Power of Yet!
- Anyone can learn math
- Impact trumps intent
- Success breeds confidence
- Strategies to help students manage frustration
- To include self-assessment and reflection

Action Plan

Assess your current efforts to Design Options for Emotional Capacity. Download the Chapter 4 Self-Assessment at https://companion.corwin.com/courses/MathWorkshopPlus.

Try It! Promoting Self-Regulation

Identify three new strategies for promoting self-regulation. Try them and monitor for differences in mindset, coping skills, and independence.

Connections

Share something you learned with a colleague or with the math community via social media using the #mathworkshopPLUS.

Chapter 5

Design Options for Perception

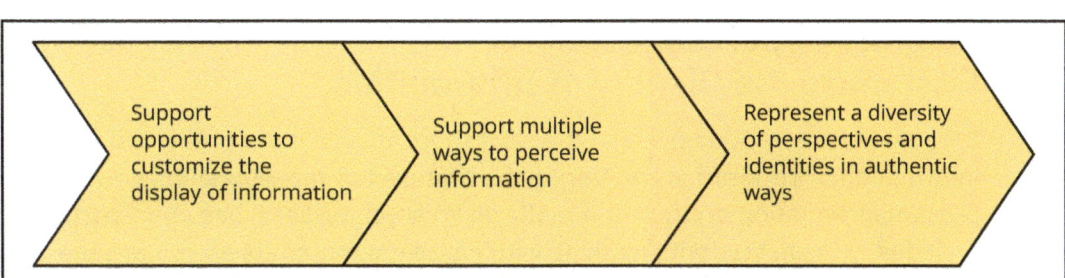

CAST (2024) points out that "learning is impossible if information is imperceptible to the learner, difficult when information is presented in formats that require extraordinary effort or assistance, and oppressive when content reinforces stereotypes or deficit thinking." Math Workshop Plus provides plenty of opportunities for customizing the representation of information. As you plan for your Math Workshop Plus learning environment, think about barriers students may encounter related to receiving and understanding information. To ensure that all learners fully understand what you are presenting, you will thoughtfully and intentionally offer the same information in different ways. This creates access. As you get comfortable, consider how you can put the controls in the learners' hands so that they can control and adjust the formats you offer. To elevate your workshop with UDL, incorporate different perspectives so students can think about things in a variety of ways.

These questions may help you stretch your thinking about *Design Options for Perception* and create a framework for Math Workshop Plus:

> 1. *Support opportunities to customize the display of information*: Does everyone have access to the information that is presented throughout the various parts of Math Workshop Plus?
> 2. *Support multiple ways to perceive information*: Do you vary access points so that students can learn through a combination of visuals, sound, touch, and more?
> 3. *Represent a diversity of perspectives and identities in authentic ways*: How are opportunities to hear and discuss different perspectives and ways of being woven throughout the learning?

Support Opportunities to Customize the Display of Information

Throughout the various parts of the workshop, students are presented with information in different ways. During the first part of the workshop, you may present information orally and visually on your whiteboard, use chart paper or slides, or have students working with hands-on materials as you engage in energizers, routines, and mini lessons. When you transition to Guided Math groups, you may use different manipulatives, visuals, or language supports, depending on the needs of the group. While students move through workstations, they may refer to anchor charts, handouts, or a digital display for information. The debrief tends to be quick, favoring an auditory approach, but sometimes you may capture the debrief on chart paper, which could include words, drawings, and even artifacts. It is important to make sure that there is a way for all students to access the information you are trying to communicate.

Mayer's 12 Principles of Multimedia Learning

Mayer's (2009) 12 Principles of Multimedia Learning are guidelines designed to optimize the effectiveness of multimedia content in promoting learning. These principles are based on Richard Mayer's cognitive theory of multimedia learning, which suggests that people learn more effectively when they can process information through both visual and auditory channels. They can be very helpful in thinking about designing options for perception. Here are the 12 principles:

1. **Coherence Principle**
2. **Signaling Principle**
3. **Redundancy Principle**
4. **Spatial Contiguity Principle**
5. **Temporal Contiguity Principle**
6. **Segmenting Principle**
7. **Pre-training Principle**
8. **Modality Principle**
9. **Multimedia Principle**
10. **Personalization Principle**
11. **Voice Principle**
12. **Image Principle**

Let's look at each one of these in more depth.

The Coherence Principle

The coherence principle notes that there shouldn't be superfluous or unnecessary information on the display because it can be distracting (irrelevant images, sounds, or text). Teachers should only include relevant content. Butterflies and ladybugs are cute, but if they detract from the focus of the math, leave them for science. Also, be sure the information being presented is aligned with the learning objectives.

The Signaling Principle

The signaling principle notes that things should be highlighted and should strategically focus students' attention on key points. Anchor charts or posters should signal exactly where the important information is being displayed. This can be done using arrows, adding a 3D effect, or circling, starring, or highlighting key points. The big idea here is that when you call attention to exactly what students should be paying attention to in the display, they are then much more likely to pay attention to it. Consider the size and position of the text. Use organizers, icons, pictures, labels, and limited text to present information. Organize information using headings and feature big ideas in bold text. Another way to focus attention on a visual display is to chunk the information into parts so students can clearly see the different ideas being discussed. Think about chunking information in the display in a way that facilitates comprehension of the ideas. Notice in Figure 5.1 how the headings support the order that students should process the information.

Figure 5.1 • Signaling Principle

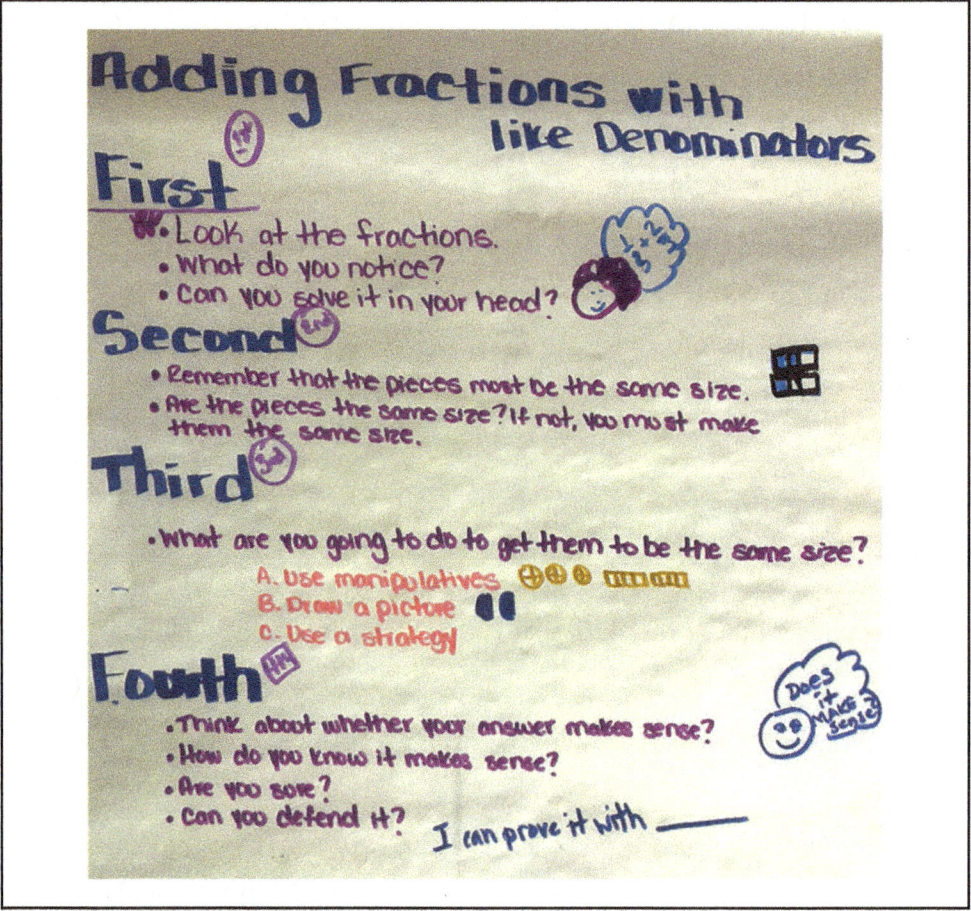

As you look at this display, answer the following questions:

- In what ways is the important information highlighted or spotlighted in this display?
- Is the information organized?
- Can the student follow along?
- Will the student remember what is being taught based on this display?

The Redundancy Principle

The redundancy principle notes that visuals should be explained with narration or text but not both. Information should not be redundant. When planning with barriers in mind, you need to consider who is in the classroom and what they need. When content is unfamiliar or difficult, you might use various methods to support comprehension and sense-making. The graphics and the text should be as close together as possible on the display so that students can make sense of the connections that are being made.

Spatial Contiguity Principle

The spatial contiguity principle notes that it is best to put the words and pictures in the graphics as physically close together as possible on the display. This helps the learner to know how information is connected. It is important to put the label next to the thing that it describes.

Temporal Contiguity Principle

The temporal contiguity principle refers to the idea that you should explain something while it is happening. This means that it's best to have corresponding words and visuals presented together if it's in some sort of video or slides instead of showing something and then talking about it or talking about it and then showing it. Presenting them simultaneously allows students to see it in action while it is being discussed.

The Segmenting Principle

The segmenting principle notes that information should be broken down into bite-sized pieces. For example, it can be very overwhelming to have an anchor chart that has 50,000 key points. Okay, we are exaggerating, but you get the idea! It's important to target key information on your anchor charts, reference sheets, or study guides. We see this oftentimes on well-intentioned math strategy charts that have four strategies on the same chart. An alternative is to break up the strategies and use illustrations to support each. Even something as simple as using bullets instead of complete sentences helps to chunk information. It is important to think about how much students are expected to process at once and how quickly they can experience cognitive overload when it's all presented together. This applies to mini lessons, small groups, and everything in between.

Figures 5.2 and 5.3 show examples of posters about adding two-digit numbers. They chunk the information in pieces to maintain focus.

Figure 5.2 • Chunking Information in Pieces on a Number Line Poster

Figure 5.3 • Chunking Information in Pieces on a Hundred Grid Poster

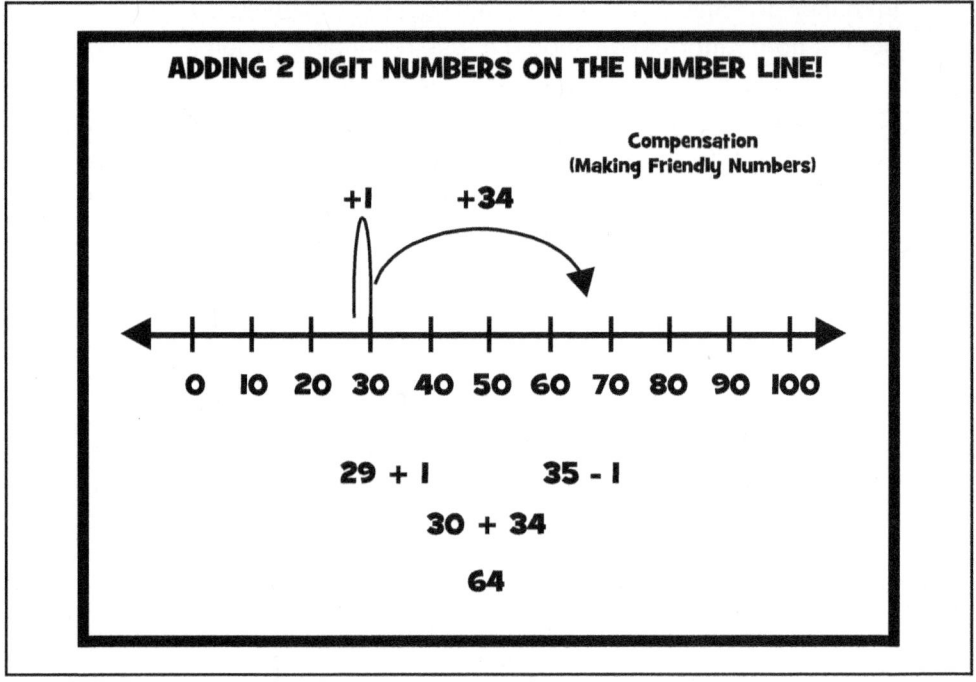

On the other hand, a poster that clumps too much information together does not follow the segmenting principal, as you can see in Figure 5.4.

Figure 5.4 • An Example of Clumping It All Together

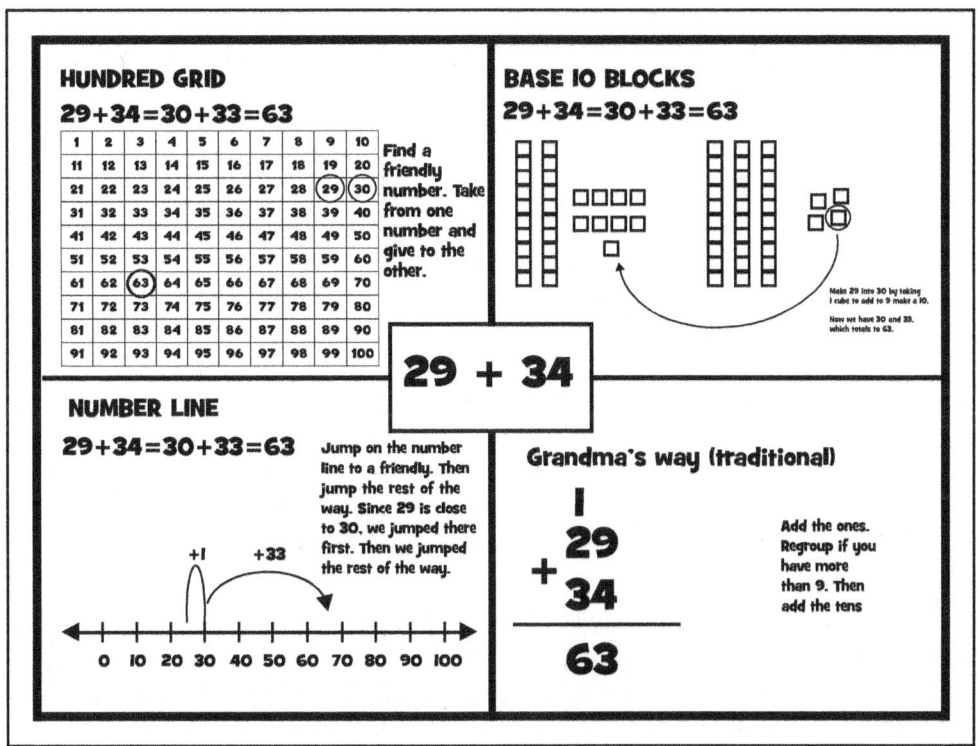

The Pre-Training Principle

The pre-training principle refers to previewing the topic, idea(s), key terms, and concepts ahead of time. In a Math Workshop Plus environment, this is like doing an anticipatory set of activities or creating a visual prior knowledge display (think K-W-L or Know/Want to Know/Want to Learn chart) to draw out and activate what students already know. Before a unit on integers, you might tap into prior knowledge by asking students where they've seen integers in the real world and creating a visual display of what they share. This allows students to begin from a place of confidence and often connects concepts in unexpected ways.

While that sounds like a lot, because it is, chances are that you noticed things you already do. As with everything in this book, focus on the areas where you can upgrade, and don't try to do everything at once. If you tend to use a lot of videos, two things that are easy to adjust but may not be on your radar are volume and the speed of the video. Volume is obviously adjustable, but it's a good idea to check in with students about the level you select. Some students have sensory issues that make them sensitive to sounds, and all students should be able to hear in a way that is comfortable. The rate of speech should also be adjustable depending on where the video is sourced. On YouTube, the speed is easily made slower or faster by changing the settings. This ensures that video narration isn't too fast or too slow. It's a great idea to teach students how to adjust this themselves so they can make videos "just right" for them whether they are in school or at home.

It's easy to have control over a video, but what about yourself—have you ever thought about the speed of your speech? Both Alison and Dr. Nicki are notoriously fast speakers, and when too much caffeine has been had, it's difficult to slow us down! We have also spoken at large conferences where microphones are mandated to ensure everyone can hear the message, and lately there has been a push to turn in captions, which is super exciting since it's a great example of UDL! But enough about us—let's get back to you.

Do you identify as loud, or are you typically soft spoken? Are you a slow, intentional speaker, or do you speak a mile a minute like us? Just as with video, when we speak too quickly some students miss the important points. If you're not sure, or even if you are, you may want to ask your students for feedback. As you know, they will tell you the truth! It is important to consider age because attention span is correlated to age, as mentioned earlier.

Activating prior knowledge doesn't have to feel dry and academic. You can simply read a picture book followed by a discussion or share a poem or song and talk about what it means to students. To foster math connections, you can show a video or take the class on a math walk and talk about what they see and know. Students can brainstorm and then talk with each other in partners and then in groups to share what they know about a topic. If you're up for it, you can even set up a graffiti wall, where students can draw and paint things that they relate to the upcoming topic, such as in Table 5.1

Table 5.1 • Example of Prompts for a Graffiti Wall About Fractions

Questions	Student Answers
What do I know about fractions? Where have I heard about them? Where do I see them in real life?	
What do we know about fractions as a group? (use words, pictures, and symbols)	
How do we feel about fractions as a group?	
Do we think they are important to know about?	

The Modality Principle

The modality principle notes people learn better when they get information in multiple modalities. You should present in visual and auditory forms rather than just one mode (e.g., all text or images). The modality principle notes that graphics should be described with narration rather than on-screen text. However, through the lens of UDL, you might do both. Mayer warns about too much in a single display because this might inadvertently cause students to experience cognitive overload.

The Multimedia Principle

The multimedia principle states that both words and graphics should be used when displaying information. The graphics should be helpful and illustrative to the discussion rather than just decorative. The big idea is that words and pictures can be very powerful when used together. The images should enhance, clarify, reinforce, explain, and describe the big idea. The picture should help to visualize the concept.

The Personalization Principle

The personalization principle notes that the presentation should be in a student-friendly, accessible style. This is important because research shows that when things are in a conversational style, in a person-to-person voice, and in a relatable context, students learn better. Think about frontloading vocabulary words. Would they be better understood with a formal dictionary definition or something more colloquial that students can understand and see? As previously mentioned, in a Math Workshop Plus environment, vocabulary is displayed and illustrated for visual accessibility.

The Voice Principle

The voice principle notes that people learn better when they hear a human voice than a machine-generated voice. Human voices feel more personal and engaging.

The Image Principle

The image principle notes that people do not necessarily learn better when the speaker's image is included in the presentation. Sometimes including the speaker's image can be distracting unless it adds value (e.g., for creating a more personal connection). These principles aim to design multimedia content that supports cognitive processing, reduces overload, and enhances learning by effectively leveraging both auditory and visual elements.

The Embodiment Principle

The embodiment principle is a related but separate principle. It is not part of the original 12 but is very relevant to classroom practice. This principle notes that having coaches and characters (photorealistic or illustrated) can help guide students through the learning process. This is very much akin to a miscounting puppet we like to use in early childhood classrooms (Figure 5.5). As the puppet makes mistakes counting, it becomes the job of the students to correct him. We also know a teacher who always highlighted "Christina's mistakes" (Exhibit 5.1). Christina was not in the class, nor was she a real person, but a fictional character the kids were always trying to help. There's something about having this kind of character in the classroom that acts out the things you don't want the kids to do. It lowers risk and is both exciting and powerful for students to look at error patterns and mistakes and then talk about what "that student" did.

Figure 5.5 • Dr. Nicki Guiding the Discussion With Murphy the Miscounting Puppet

Exhibit 5.1 • Christina's Mistake

Christina solved the problem this way:

78

57

1515

What was her mistake? How would you tell her to fix it?

Connections to SEL

Self-Awareness

In a Math Workshop Plus classroom, students are taught to constantly evaluate their comprehension of what is being taught. Through modeling your own metacognition, you can support students in building their own self-awareness. Our goal is to build a supportive learning environment where students can express if they are getting the information and comprehending the information. You can promote this self-awareness by asking questions and having students engage in activities where they must reflect and think about what they have just learned.

> **Equity Check**
>
>
>
> Opportunity for All to Learn
>
> If curriculum is going to be accessible to all students, you need to plan with an equity lens. Equity as accessibility for all learners shapes how we think about not only *what* we are teaching but *how* the students are accessing the information. In our practice, we don't see this emphasized enough because due to pacing guides and mandated testing and other pressures, we tend to focus more on the what than the how. Take a moment to consider that and decide where you are. Do you think equally about the how? If not, how might thinking equally about the how result in adjustments you could make that would result in more equitable access? We all must remind ourselves that all students deserve an "equitable opportunity to learn . . . regardless of characteristics and identified needs . . . and receive" whatever supports necessary to reach academic excellence (IDRA, 2020, Goal 4).

Support Multiple Ways to Perceive Information

As we just discussed, you probably use a variety of "images, graphics, animations, videos, [and] text" to present information. We have discussed some ways to make these modalities as accessible as possible, but you may be starting to wonder how you will possibly add all of these supports without working around the clock! It is totally normal to feel that way, but don't worry because once you get the hang of this, it becomes your way of doing business and doesn't feel like something extra. In the meantime, to assist you in planning a universally designed Math Workshop, check out the sample accessibility notes we created to get you started. Table 5.2 outlines strategies for each component of your workshop. Remember that you need to look to your learners to see which supports you will include, and trying to do too much at once won't work. Be strategic about what you add and watch to see what changes as a result. You will add more as you go, and this is only meant to offer a structure you can use and to reinforce some of the ideas we have reviewed.

Table 5.2 • Math Workshop Plus Accessibility Notes

Lesson Components	Accessibility Notes
Energizers/ Routines	• Use concrete materials (cubes, coins, teddy bears) • Include visual aids (number lines, number grids, number frames) • Provide descriptions (text or spoken) for images, graphics, videos, or animations • Use text guides • Provide visual diagrams and prompts
Mini Lesson	• Encourage discourse • Add visuals • Include relevant, concrete examples • Act out problems • Provide descriptions (text or spoken) for all images, graphics, videos, or animations • Ensure on-demand access to math tools and manipulatives
Guided Math	• Encourage discourse • Add visuals • Include relevant, concrete examples • Act out problems • Provide descriptions (text or spoken) for all images, graphics, videos, or animations • Ensure on-demand access to math tools and manipulatives • Make graphic organizers accessible
Workstations	• Use recordable buttons or a tape player to record directions and important information • Ensure on-demand access to math tools and manipulatives • Add visuals to directions • Design strategic partnerships to support accessibility (read the problem, etc.) • Pre-teach workstation activities in small groups • Offer varied options for accountability • Make headphones available
Debrief	• Use concrete material (cubes, coins, teddy bears) • Include visual aids (number lines, number grids, etc.) • Turn on speech-to-text on your screen • Allow students to share their learning using a variety of modalities (talk, draw, model with manipulatives, etc.)

 Download the Math Workshop Plus Accessibility Notes at https://companion.corwin.com/courses/MathWorkshopPlus.

Creating Accessible Tables and Charts

Tables and charts often present challenges for students. To support comprehension of information presented in tables and charts in your Math Workshop Plus, think about how to make them more accessible. What attributes can you adjust so that all users can understand and interact with the data? In addition to the ideas already explored, Exhibit 5.2 offers some ideas to upgrade your tables and charts for accessibility.

Exhibit 5.2 • Accessibility Notes for Paper Tables and Charts

Use Descriptive Titles and Labels
- Add clear and concise chart or table titles.
- On charts, label axes.
- On tables, ensure there are header rows for columns and rows.
- Avoid abbreviations unless they are commonly understood.
- Consider the use of pictures or concrete tools.

Provide Alt Text or Descriptions
- Write descriptive alt texts for charts, summarizing the key points.
- For complex charts, break information down.

Color Contrast and Differentiation
- Use high-contrast colors to make charts readable for people with visual impairments or color-blindness.
- Use a variety of approaches to convey meaning: patterns, textures, and labels to make things pop.

Accessible Fonts and Sizes
- Use clear, easy-to-read fonts that are big enough to read (12 point or larger).
- Make sure text is readable on all parts of the table or chart.

Zoom-In Side Chart
- Have a well-organized "decoder chart" on the side to help to interpret the chart visually.
- Be sure the table and zoom-in chart match.

Avoid Too Much Information
- Keep it simple.
- Keep it spaced (not crowded).

Solicit Feedback About Your Charts/Tables
- Ask colleagues to look at the chart and share suggestions.
- Have students discuss the accessibility and give feedback.

Source: Missouri State Admin, 2017; Library City University of Seattle, 2024.

A Deeper Look at Sound in the Classroom

There are some very specific questions that we can ask about sound in terms of our classroom.

- What is the soundtrack of our classroom? Is it audible?
- Who speaks?
- Who doesn't?
- Who gets heard?

What accommodations do you typically add to ensure that everybody gets access to the information being shared and received auditorily? As you grow your UDL lens, you will likely be less inclined to use singular modalities, as they often exclude certain students. In this case, an opportunity gap could be created for students who have limited hearing or who struggle with auditory processing. These students will need extra time to process what they hear and therefore may be a step behind you. We've seen students reprimanded for not listening when in reality they couldn't keep up with the pace of the auditory input. Alison's daughter has an auditory processing disorder, so she is particularly attuned to the extraordinary energy it takes for these students to attend in this modality and how they are often perceived as off task or not listening.

A similar dynamic exists for students who experience challenges with memory. When a high volume of auditory information is coming at them, how will they anchor it so that it can be processed, understood, and retained? How will you know if these students understand what was said, what it meant, and what are they expected to do with that information? Can you think of a student right now who fits this description? As you look at this student through the lens of UDL and equity, what is currently missing from the learning environment? What could be added to unlock the ability of this student to thrive?

Math Workshop Plus will continually prompt you to consider these and other questions. As you plan lessons with these questions in mind, you will design and build structures and supports to mitigate them before you begin teaching. For example, adding a listening protocol with a graphic organizer can help students keep track of important information when information is presented orally. You might remind students about choosing to use the protocol to capture notes during a class discussion where visuals aren't present. The protocol might look like the one in Figure 5.6 for a discussion on rounding. When you create a protocol, match it with the flow of the lesson and the information you know you will hit. Making this available to all students allows for choice and helps students discover what works for them.

Figure 5.6 • Listening Protocol

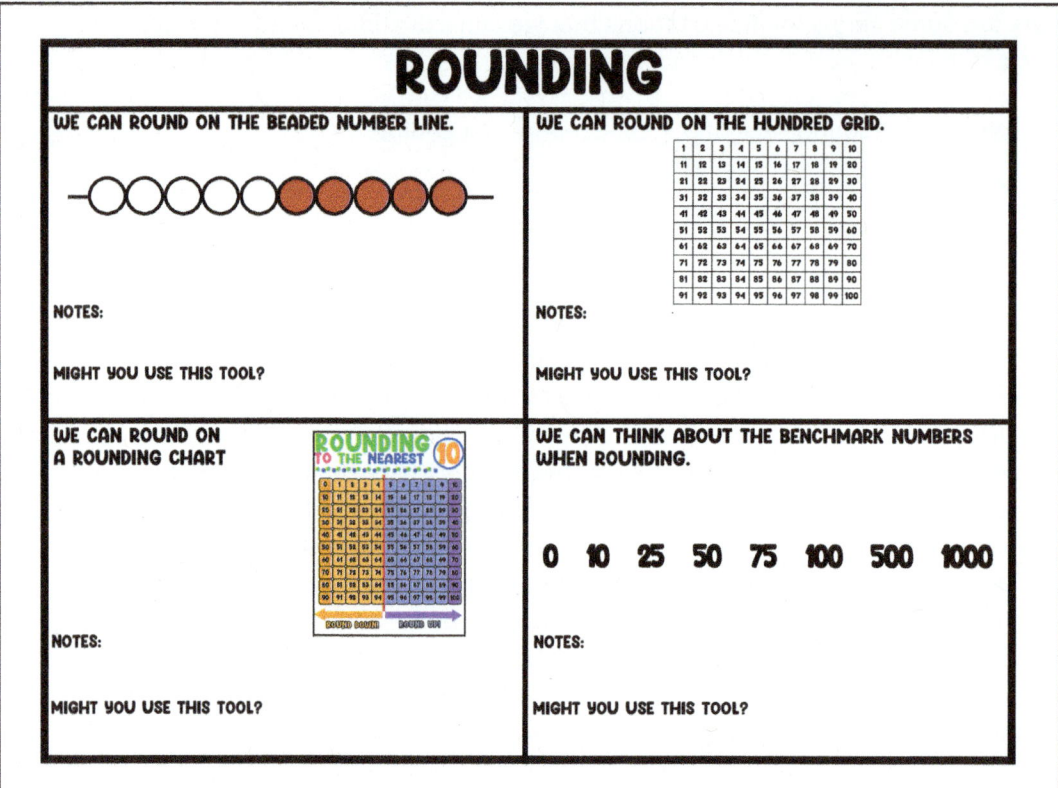

Verbal Charts to Accompany Discussions

In a vibrant Math Workshop, students are collaborating, talking, and listening to one another. Sometimes information is shared so quickly that students don't have time to comprehend and grasp the ideas or jot down notes. Some students have trouble orally processing information as well as retaining things in their working memory. Let's look at some examples that support discussion using different levels of visuals, text, and prompts.

One way to start is with a Partner Reflection Checklist. First, the teacher gives the verbal instructions:

- Sit with your partner.
- Discuss what you did.
- Take turns talking.
- Come back and share with the group.

Students then use a Partner Reflection Checklist that covers the following tasks:

- ☐ I talked about my work.
- ☐ My partner talked about their work.
- ☐ We listened to each other.
- ☐ We showed each other our work.
- ☐ We shared with the whole class.

You might have a poster or handout that leverages the accessibility guidelines we've talked about and looks like Figure 5.7.

Figure 5.7 • Partner Reflection Checklist

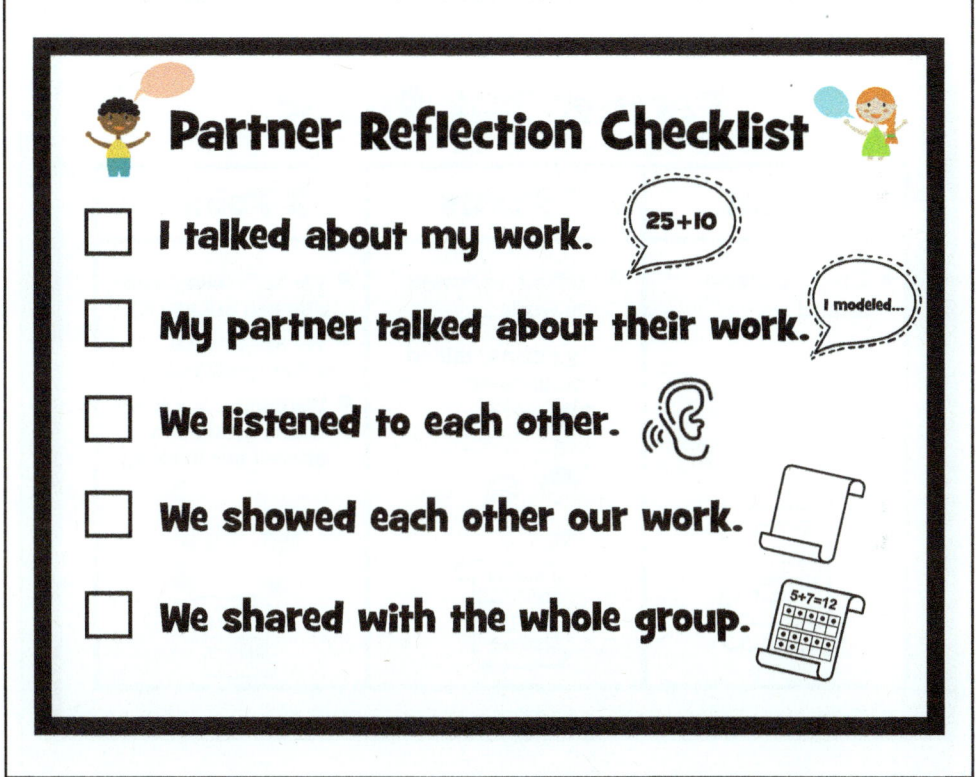

Source: Newton, 2024.

 Download this Partner Reflection Checklist at https://companion.corwin.com/courses/MathWorkshopPlus.

Another tool is a Partner Talk Rubric, as shown in Figure 5.8. In this case, you share a 3-point rubric.

Students mark themselves for the following:

1 point = Only one person talked.

2 points = I talked. I showed my work *and* my partner talked and showed their work.

3 points = We both talked and showed our work *and* we asked each other questions *and* we came back to the group and shared our thinking.

Figure 5.8 • **Partner Talk Rubric**

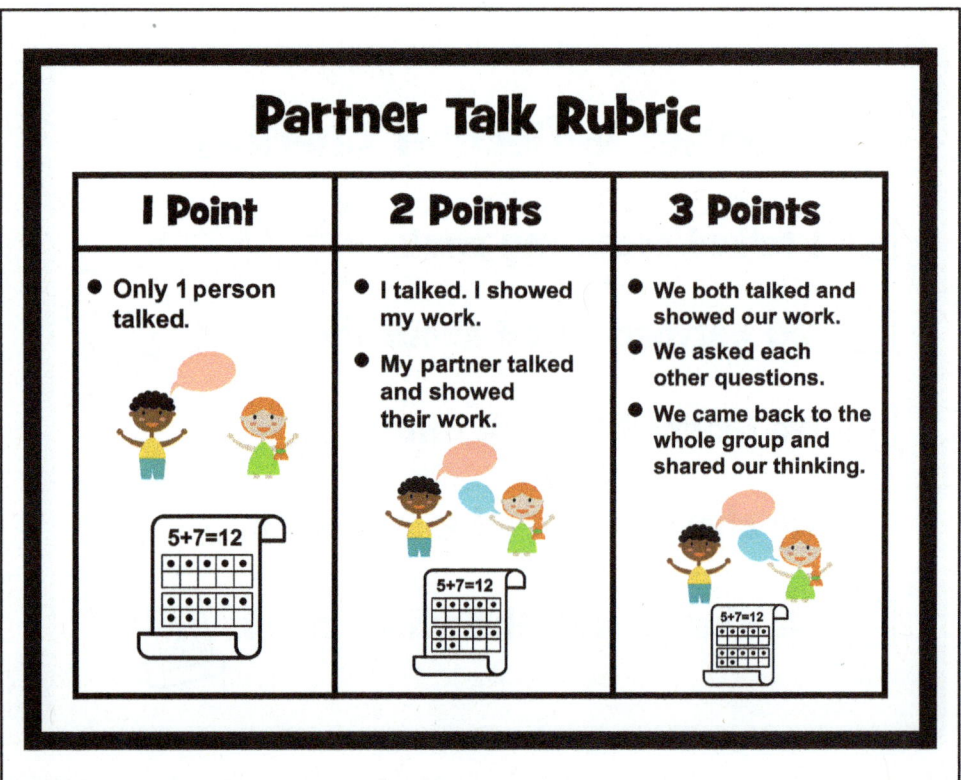

Source: Newton, 2024.

Download the Partner Talk Rubric at https://companion.corwin.com/courses/MathWorkshopPlus.

Captioning Slides and Videos

Teaching has changed so much since we started our careers over 30 years ago. The constant advances in technology make it much easier to bring UDL elements into your instruction. For example, did you know that you can easily turn on captions in Google Slides and PowerPoint? This feature automatically transcribes any audio input, whether it's you speaking, students speaking, or a video playing. These captions can be moved to the top or bottom of the screen, and the text size is also adjustable. At the conclusion of the slideshow, you can even download a transcript!

When watching videos, you can also typically access a written transcript. Students (or you) can annotate the transcript to support comprehension. To have students annotate slides as you are presenting them, print them in the format that leaves space for notes next to each slide. This is another way for students to see and be able to follow and process the information in a logical manner. For example, Figure 5.9 shows a PowerPoint slide for subtracting fractions that shows a closed caption. You can see in the body of the slide a graphic organizer for showing the set model, the area model, and a linear model. Exhibit 5.3 shows a graphic organizer you can create as notes next to the slide for students to annotate.

Figure 5.9 • A PowerPoint Slide for Subtracting Fractions With Captions

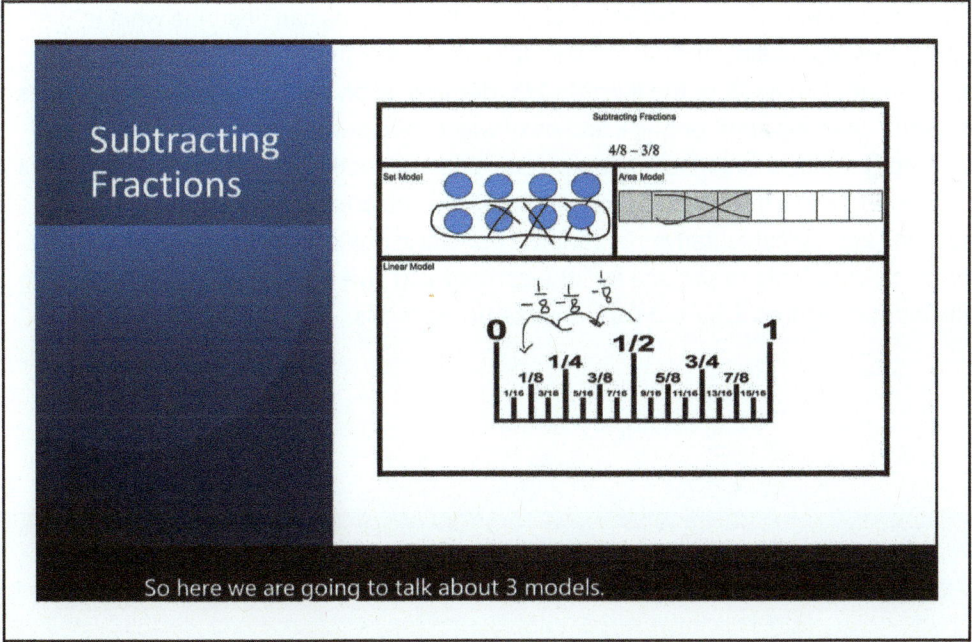

Exhibit 5.3 • A Graphic Organizer for Subtracting Fractions That Mirrors the PowerPoint Slide

Subtracting Fractions	
Set Model	**Area Model**
Linear Model	

Alternatives for Visual Information

We have spent a great deal of time thinking about how to support language by adding visuals, whether with clip art, photos, videos, posters, or captions, but what if the barrier students have is with visual input? How can you create access for students who have challenges processing visual information? Features like audio notes can help support understanding of visual content. These notes can describe what is seen in the visual field, such as a cake cut into 6 square pieces with 2 missing as a representation of 6 – 2, or the fraction 1/3, depending on the grade. Adding QR codes to anchor charts is another quick and easy way to add audio input to accompany the visual content. In addition to audio, you can also add tactile resources to aid students in accessing visual information. For example, you can create anchor charts by attaching actual manipulatives or various real-world objects (realia), such as coins. This allows students to see and feel the content and give them enough information to talk about it. Figures 5.10 and 5.11 are examples of this.

Figure 5.10 • Anchor Chart With Realia

Figure 5.11 • Anchor Chart With a QR Code for an Audio File

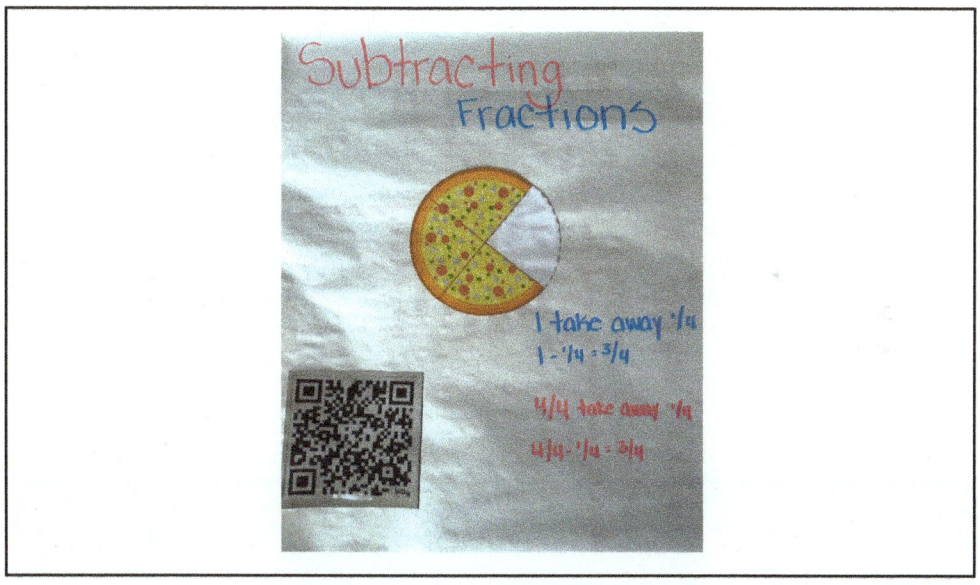

There are countless ways to add accessibility to your classroom, and so many are free! If you feel intimidated by them, try them for things outside of your classroom before using them with students. If you stop and think about it, maybe you already do. Do you use speech-to-text in your car or on your phone?

Do you enlarge the font on things so you can easily see without your glasses? Do you listen to audiobooks or watch videos with the sound off and captions on? These things have become normal within our daily lives and can easily enhance our learning environments. In some cases, it's as easy as turning on an accessibility feature. In situations where more may be needed, such as having Braille books available or sound fields and microphones for students with hearing impairments, you may need to consult with someone who has specialized knowledge.

The big idea with this component of UDL is "to provide content that, when presented to the user, conveys essentially the same function or purpose as auditory or visual content" (Letourneau & Freed, 2000, Guideline 1). What's important is to ask yourself, what am I about to show? Who has access? Who doesn't? What do I need to do to make sure everybody gets this information and can use it? Once you get the hang of using these features, you will probably wonder why you haven't used them sooner!

Connections to SEL

Multiple Ways to Access Information

When information is presented in multiple ways, students can choose how they best want to get that information. For example, when you add a QR code to an anchor chart, you give students the opportunity to scan it and hear about the information on the chart, as well as extend the conversation. While some students may elect to do this, others might prefer a tactile chart that offers them the chance to physically touch and feel the different manipulatives on the chart. All these things involve being self-aware, managing one's own learning, and making decisions about how one wants to receive the information.

Equity Check

Systems of Delivery

What used to be, no longer is. We live in the digital age, and an equitable classroom environment reflects that. Researchers have noted that many of the teaching methods designed for the industrial age are still dominant in

> today's classrooms. How can this be? More importantly, what will we do about it? Systems of delivery from anchor charts to computers must all consider accessibility. Through an equity lens, we must ensure that all students can access, use, and manipulate the tools and resources that they are given. Part of equity is empowering students to say when something isn't working for them. They should be able to communicate that they need to see or hear information again or in a different way. Teaching students to be self-advocates is important both in and outside of school.

Represent a Diversity of Perspectives and Identities in Authentic Ways

One of our favorite things about Math Workshop Plus is that it holds space for a variety of ideas to be expressed, and it encourages diversity of thought and ways of knowing in our classrooms. As you design your tasks, look for opportunities to create "mirrors" where students can see reflections of themselves and "windows" or "sliding glass doors" where they can "walk through" and see and experience the perspectives of others (Bishop, 1990). If you're not sure how to do this, it can be as simple as exploring how people from different countries solve the same math problem. For example, in the United States we favor long division, but in many countries, they focus on short division. When new students join your school community, whether from another state or another country, encourage them to share their ways of knowing and doing math. Not only will this give you insight to other instructional approaches used, but it will respect the diversity the student brings and honor their way of thinking.

◇◇◇◇◇◇◇◇◇◇◇◇◇◇◇◇◇◇◇◇◇◇◇◇◇◇◇◇

One of our favorite things about Math Workshop Plus is that it holds space for a variety of ideas to be expressed, and it encourages diversity of thought and ways of knowing in our classrooms.

Students who join us from other countries not only have to assimilate to our environment but must use a language that is often not comfortable or familiar to communicate their thinking and ideas. One way to make your classroom feel welcoming and inclusive is by welcoming translanguaging. Translanguaging is "the ability to move fluidly between languages" in the classroom. In a translanguaging classroom, students think about things, process information and come to know in "multiple languages simultaneously, and use their home language as a vehicle to learn academic English" (Najarro, 2023). For example, Dr. Nicki was recently in a classroom where she was working on division story problems. She told a story followed by a question.

She told the students that she went into the forest and saw 18 legs hanging from trees. Then she asked how many bats were there. Everyone was working, and as she visited each table, when she got to a table where students had just come from Mexico, they began to translanguage. First, they established what "bats" were in Spanish. Then they discussed the problem in Spanish before talking about it in English. Every table got a chance to go around at the end and share their models and thinking. All students felt that their ideas were included and valued, and because of this, they felt a sense of belonging to the community of learners.

If you've ever felt like you didn't belong, you know how important it is to get this right. Think about how you can create a sense of belonging in your math class using posters and anchor charts featuring multicultural children, including the children from the classroom. During the First 20 Days, we talk a great deal about having a growth mindset and persevering when problems get tricky. Including posters of different children using growth mindset language can help students see themselves represented. Diversity is more than just ethnicity, so zoom out to be sure to that diversity includes neurodiversity, physical diversity, and other differences that may not be as visible. Figures 5.12, 5.13, 5.14, 5.15, and 5.16 will give you some ideas.

Figure 5.12 • Poster: "I'm Learning More Every Day!"

Figure 5.13 • Poster: "I Believe in Myself!"

Figure 5.14 • Poster: "I Am Smart!"

Figure 5.15 • Poster: "I Do My Best!"

Figure 5.16 • Poster: "I Work Hard!"

Connections to SEL

Importance of Belonging

Students learn to respect, honor, and get along with different people as they are given the opportunity to do so. Throughout Math Workshop Plus, we provide students with opportunities to get to know about others and to work with people different from themselves. These opportunities help students to develop strong relationships and to live and embrace the differences among us. These opportunities also give students the opportunity to celebrate their differences and engage in a community where everyone is accepted and appreciated for who they are!

Equity Check

Importance of Belonging

When students feel that they are cherished and liked, we build equity in the form of inclusivity and belonging. A friend of ours told us that her son came home and told her he had a good day at school. His mom asked why. He said his friends played with him and they liked him. This tells it all. He didn't say it was a good day because he passed his spelling test or got 100 on his math benchmark. He said it was a good day because he played, and other kids liked him. His good day was based on relationships. That is where it all begins, and you have the power to set the stage for these relationships to thrive.

Say Bye-Bye to Barriers When You . . .

- Make sure the same information is presented in more than one way
- Make sure that displays are organized in a way that spotlights what is important
- Consider the speed and pace of the auditory information
- Make sure the visual information is accessible to all students
- Think about ways to add audio to visual information
- Put the actual manipulatives on anchor charts
- Integrate diverse perspectives and identities throughout the curriculum

 **Action Plan**

Assess your current success with Designing Options for Perception. Download the Chapter 5 Self-Assessment at https://companion.corwin.com/courses/MathWorkshopPlus.

 Try It! Enhancing Information Representation

Choose a new strategy to enhance how you represent information for your kids. Try it.

 Connections

Post an artifact of something that you have done to address representing information on social media and tag it with #mathworkshoplus.

◇◇◇◇◇◇◇◇◇◇◇◇◇◇◇◇◇
Chapter 6

Design Options for Language & Symbols

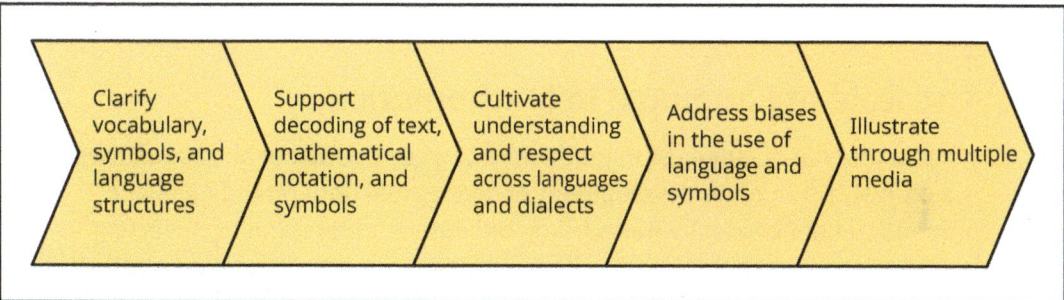

When thinking of math, people often envision numbers and calculations. But mathematics is more than naked numbers and rote computation. Math involves thinking, reasoning, understanding, and explaining, which require students to have facility with language. Have you ever lamented the fact that a math textbook or state assessment was language heavy, leaving you to question if student performance accurately represents math proficiency? When you can't determine if struggles are due to math or language, it leads to uncertainty about what interventions are needed.

If we view math through the narrow lens of procedural fluency, we may overlook the role of language or discount the need to explicitly teach the vocabulary and symbols that impact reasoning, strategic competence, and conceptual understanding. Math is a tool to make sense of the world (Su, 2017), and language plays a vital role in making sense of problems. Take a moment to really think about the implications of this outside of school, where knowing how to interpret and understand information affects every aspect of life. As you look at your classroom through the lens of *Design Options for Language & Symbols*, consider these questions to reflect how you currently approach language, vocabulary, and symbols:

> 1. *Clarify vocabulary, symbols, and language structures*: How is academic vocabulary supported?
>
> 2. *Support decoding of text, mathematical notation, and symbols*: What visuals are present to support decoding?
>
> 3. *Cultivate understanding and respect across languages and dialects*: How might highlighting common roots and prefixes support understanding?
>
> 4. *Address biases in the use of language and symbols*: Is language required to demonstrate proficiency?
>
> 5. *Illustrate through multiple media*: How is technology maximized to enhance comprehension?

The Role of Language in Mathematics

What is the role of language in math? As noted by Driscoll et al. (2016), language is a crucial factor in the development of conceptual understanding. Since procedural fluency ideally flows from conceptual understanding, inserting language supports helps ensure the connection is made. Language is also a primary driver in one's ability to construct viable arguments and critique the reasoning of others, one of the Standards of Mathematical Practice (SMP) outlined in the Common Core State Standards (CCSS, 2010) and most adopted state standards.

Word problems immediately come to mind when thinking about language in math. Word problems confront students with academic language they must understand to successfully solve. As defined by Schleppegrell (2004), *academic language* refers to the linguistic expectations of students to learn, speak, read and write about academic subjects such as mathematics. When students lack facility with academic vocabulary, it is challenging to solve word problems. In the classroom, this may look like students calculating, often accurately, but lacking awareness that the operations do not match the context of the problem.

What does it look like to Design Options for Language & Symbols? First, focus on what you know about language and comprehension. If you teach literacy, reflect on how you approach comprehension in reading and writing and how those strategies compare to those used in math. For example, are students asked what is happening in the "story" of the problem, or are they told to circle or underline signal words and key information to find an answer? When we revisit our frustration about "language heavy" problems, we must ask ourselves how we are preparing students to make sense of language. What small shifts in the learning environment could empower students—all students, including those with limited language proficiency—to participate and accurately show what they know and are capable of (CAST, 2020)?

Connections to SEL

The Importance of Language Supports

Wakefield (2000) states that students who are pushed to use math language they don't yet understand may experience math anxiety. Research indicates a strong correlation between math achievement and math anxiety (Barroso et al., 2021), and students who report higher levels of math anxiety typically have lower achievement in math. As you implement strategies to support language, think of them as proactive moves to prevent this anxiety from developing.

Clarify Vocabulary, Symbols, and Language Structures

Think about the vocabulary and symbols students encounter as they learn math throughout their education. Math has its own language, and if you don't know it, you are excluded from the conversation, the thinking, and the opportunities to make conjectures and discover connections (Wakefield, 2000). Vocabulary is particularly challenging for multilingual learners. Rubenstein and Thompson (2002) outline the following 11 issues (yes, *11*) that typically cause confusion as students encounter language as they are acquiring it:

1. The same words can have different meanings in math than in everyday English. (There are 12 inches in a *foot*—but I stepped on a bug with my *foot*.)

2. The same words can have similar meanings in math and everyday English, but have more precise mathematical interpretations. (*Difference* means the answer to a subtraction problem—that is, the *difference* between two objects.)

3. Some words are math specific and only found in math contexts. (*numerator, quotient*)

4. Some math words have more than one meaning. (In math, *square* meaning a shape; *square* meaning a number times itself; *round* as an adjective describing a shape; *round* as a verb, as in round to the nearest ten.)

5. Some words are shared across content areas but mean different things in each. (*Variable* in math representing the unknown versus independent or dependent *variable* in science.)

6. There are some math words that are homonyms to everyday English words. (*pi* vs. *pie*; *sum* vs. *some*)

7. Some math words are related, but their meanings are often confused. (*hundreds* vs. *hundredths*; *factor* vs. *multiple*)

8. A single English word that is used in math and everyday English with different meanings, such as *table*, may translate into two separate words in another language such as Spanish. (table used to eat dinner—*mesa*—vs. table used for math data—*tabla*)

9. Spelling and usage in English may be inconsistent. (*Four* contains the letter "u," but *forty* doesn't; teen numbers don't follow the same convention—*eleven* and *twelve* don't use the -teen suffix as *thirteen*, *fourteen*, *fifteen*, and so on do.)

10. Some math concepts are expressed orally in more than one way. (*quarter-past* three, 3:15; *one-quarter, one-fourth*)

11. Students may know and use a colloquial term as if it were a precise math term. (*corner* vs. *vertex*; *diamond* vs. *rhombus*)

We don't know about you, but we saw the complexities of language through a whole new light after diving into that list. It's no wonder students have trouble! While it may seem like a small thing to clarify vocabulary and symbols, it can clearly make a big difference. In Math Workshop Plus, we use UDL to create a learning environment that helps students construct meaning from words, symbols, and numbers using different representations. In this chapter, we will identify opportunities and strategies to do the following:

- Pre-teach vocabulary and symbols with an eye on making connections to prior knowledge

- Provide graphic representations paired with text and symbols.

- Highlight how terms, expressions, and equations that are complex are composed of simpler words or symbols
- Provide access to resources that support comprehension (definitions, visuals, etc.)
- Offer dynamic ways to support language acquisition and comprehension

Adapted from CAST, 2024

Classroom as a Tool Kit

So much of the success of Math Workshop Plus comes from the classroom learning environment (CLE). The CLE can support independence, build confidence, and invite students to own their learning, or it can do the opposite (Schoenfeld, 2020). When we view our classroom as a tool kit, we use our walls, shelves, technology, and math resources such as manipulatives, protractors, anchor charts, and reference materials to increase accessibility. Let's explore some of these tools and how to use them to clarify vocabulary, symbols, and language structures.

Equity Check

Do All Classrooms Have What Students Need?

If students in different classrooms do not have access to the same tools and resources, this is an equity issue that must be addressed. If this applies to you, raise the issue immediately and look for ways to correct it, such as taking inventory and redistributing resources, seeking local grants, or asking admin to purchase additional resources.

Language Frames/Sentence Stems

Sentence stems are a great tool to promote language use, support accountability for explaining reasoning, and level the playing field so all students can enter the conversation. Typically, they are written on sentence strips to promote discourse and help students communicate their thinking. Language frames are similar, except they are written in more of a cloze format, so students need only fill in the blanks. Writing language frames on table tents is a great way to support discussion between two students. Student A uses the prompts on their side of the tent, and Student B uses the prompts on the other side. This setup supports a back-and-forth dialogue even when students struggle with language. The frames also help keep the discussion focused and on target. See Figures 6.1, 6.2, and 6.3 for examples of sentence stems, language frames, and table tent language frames. Model them so students become familiar with the pattern of the words and understand how they are expected to use them.

Figure 6.1 • Examples of Sentence Stems for Supporting Student Discourse

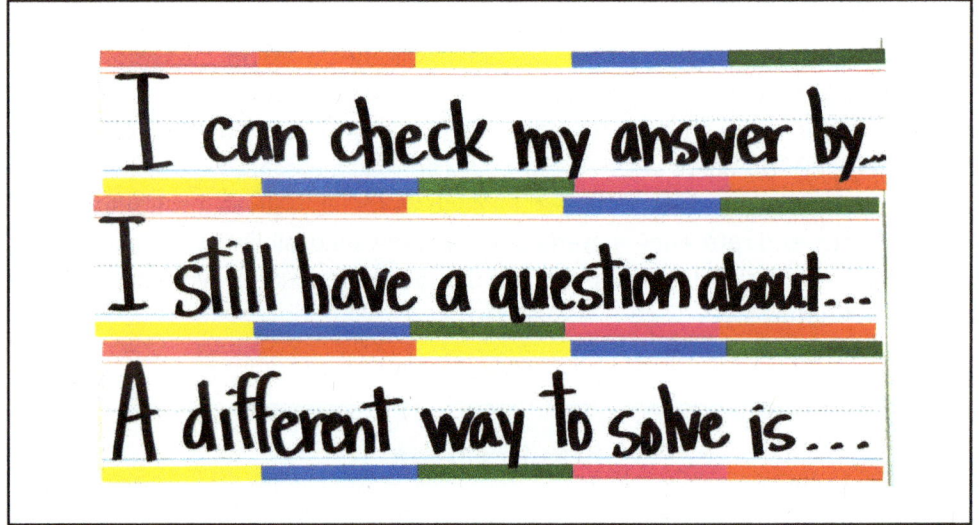

Figure 6.2 • Examples of Language Frames for Supporting Student Discourse

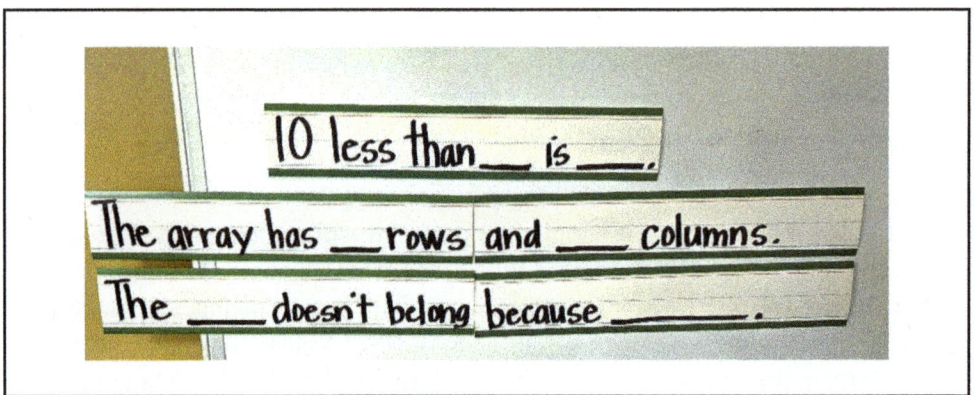

Figure 6.3 • Example of Tented Language Frame to Support Discussion Between Partners

Visual Vocabulary Cards

Visual vocabulary cards are self-explanatory, and there are many free sets available online. These cards feature a vocabulary word accompanied by a visual representation. Sometimes the cards have cleverly embedded representations that promote understanding even further, something like you see in Figure 6.4 where the word *perimeter* outlines the rectangle, highlighting the R-I-M in each word to help students associate the meaning of the word with "rim."

Figure 6.4 • A Visual Vocabulary Card for the Word *Perimeter*

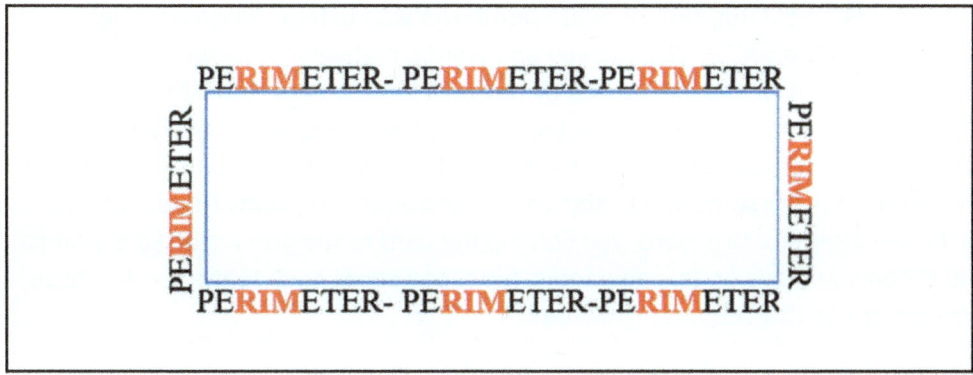

Others may include a written definition like you see in Figure 6.5, in which the word *trapezoid* is accompanied by a definition and a visual representation:

Figure 6.5 • A Visual Vocabulary Card With a Definition for the Word *Trapezoid*

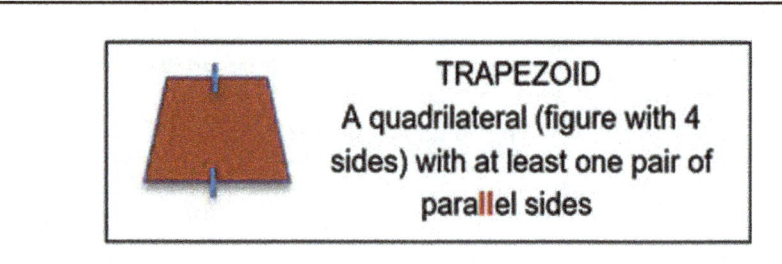

Visual vocabulary cards are ideal for classroom word walls or interactive notebooks. Use them to preview vocabulary or to add support throughout a unit of study. The cards may be introduced during a warm-up, mini lesson, or Guided Math group and may be revisited in workstations.

Word Walls

A word wall is what it sounds like: a place on the wall where you post vocabulary terms you want students to reference and remember. If you want your word wall to be interactive, be strategic in how you construct it—for example, at the beginning of the year you might include vocabulary from previous grades. As part of your classroom orientation, note that students may be familiar with these words, and if not, they should refer to the wall to learn them and their meanings. As the wall grows and you add words, highlight those words so that they stand out. You can do this by having a specific space for them, like a separate column or designated spot for "active" words. Or you can also print your cards on double-sided card stock, using black ink on the front and green on the back for the same word. Add Velcro, and when you want students to pay particular attention to a word, you can flip the card to the green side so it stands out. When you move on to a new topic, flip those cards back to the black side so they are still visible but not highlighted.

To keep it simple, cluster cards by domain, topic, or chapter. Figures 6.6 and 6.7 show you examples for elementary and middle school.

Figure 6.6 • Elementary Example of a Word Wall

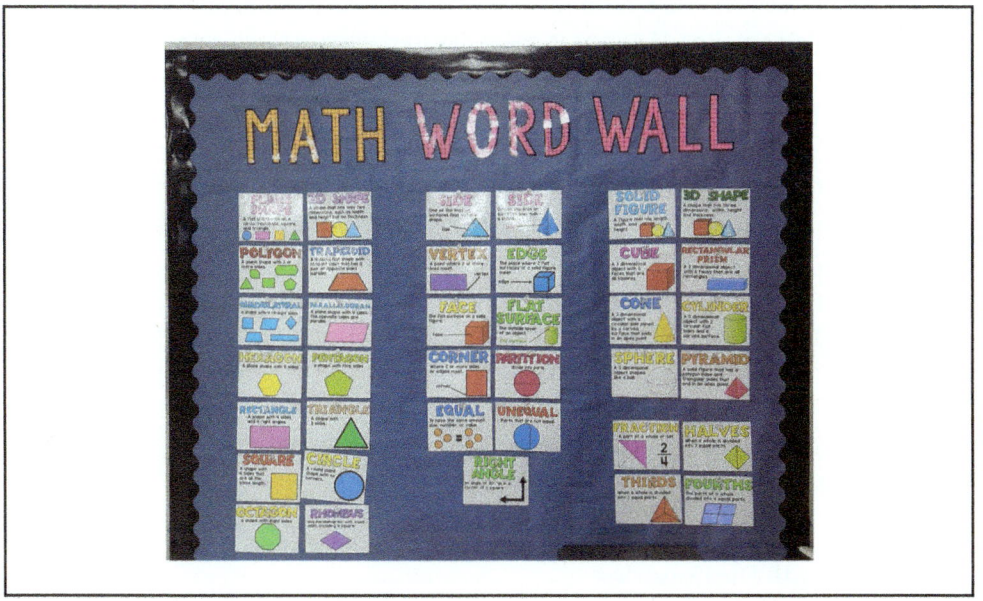

Figure 6.7 • Middle School Example of a Word Wall

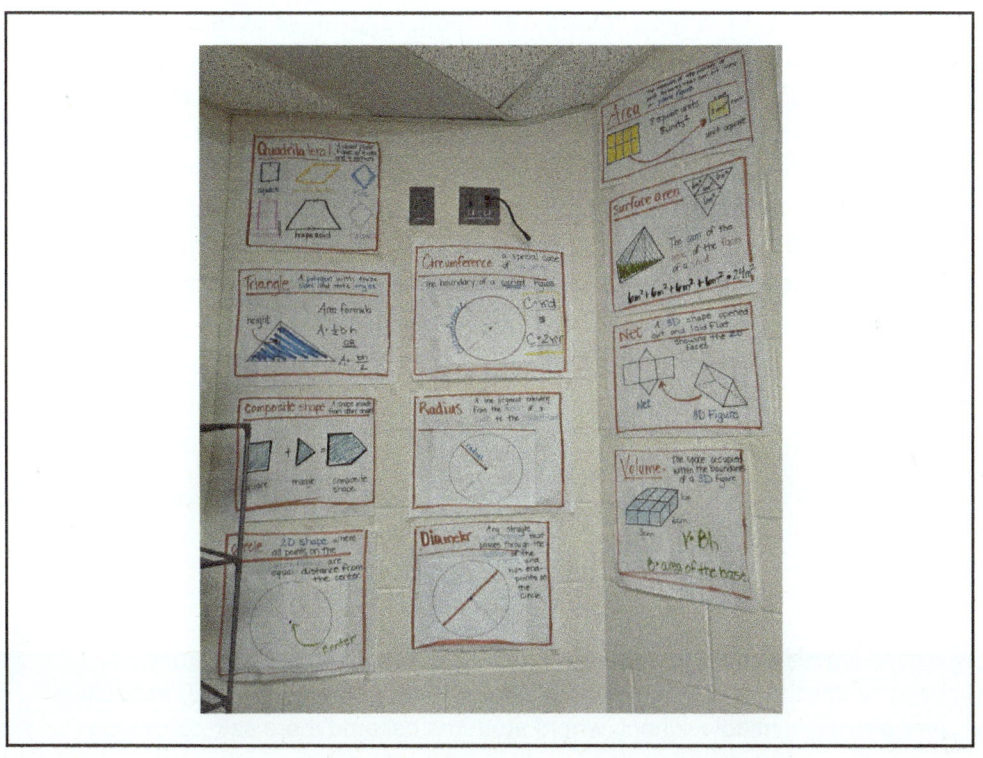

Chapter 6 • Design Options for Language & Symbols

Depending on your available wall or bulletin board space, this type of display may or may not be a viable option for you. If you lack a large space, consider making the cards smaller. If space is really limited, incorporate a small version of visual vocabulary within student reference folders, a section in a journal or interactive notebook, or on a ring.

The bottom line is having a space for visual vocabulary cards is better than not having them at all, and there really is no right or wrong way. Whatever works for you is the right way. Whether you change them out often, leave them up all year long, or use one of the strategies mentioned, making visual vocabulary a part of your repertoire will diminish the language barrier. Remember to refer to the word wall during instruction, and your students will do the same. This is a quick and easy UDL addition that will help students make sense of math language and symbols. As stated earlier, many of these cards are available for free, so you're a quick internet search away from having them in your classroom!

Illustrated Math Dictionaries

Math dictionaries are not new, but the versions with color illustrations deliver the accessibility we are after. The pictures help fill in gaps for those unable to decode written words. When these are available, students have on-demand access to any word and definition, regardless of their fluency with language. The dictionary removes a barrier so students can access the math, and there's nothing extra for you to prepare or plan for! If these sound great but cost prohibitive, there are free digital options. The one that we use from Maths Is Fun (note the "s" in *maths*) is available free online. Currently, it doesn't read to students, but many words include YouTube videos so captions may be turned on and translated. When modeled and linked to a page that students freely access, we find it is used often, and math vocabulary becomes much stronger.

Connections to SEL

Normalizing Tools for All

When something makes students stand out from their peers, they may become uncomfortable, which can negatively affect learning. This is especially pronounced in middle school, where students become more self-conscious and concerned about what their peers think of them (Bishop & Pflaum, 2005). Normalizing tools as part of learning for all, not just those perceived as needing them, increases the likelihood that students will utilize them. To support this, explicitly teach how to use tools during the First 20 Days of school and set the expectation that students use them at their discretion.

Vocabulary Workstations

A great place to fortify vocabulary is through Math Workstations. Students can interact with math vocabulary in several ways to acquire the necessary language of math in their grade level. You may be familiar with the Frayer model, and while we recommend it, we want to share some additional ideas. The examples in Exhibits 6.1 and 6.2 have been adapted from other content areas.

Exhibit 6.1 • Vocabulary Workstation—Group and Label Station

Group and label vocabulary workstation

Type of Workstation: Partner or individual

Materials: Math vocabulary cards

Directions:
1. Spread all word cards face up on your work area.
2. Carefully read each word.
3. Group the words however you like. Just make sure you can justify why you grouped them the way you did.
4. Use a blank card to write a label for each group.
5. Create a word web to show how you grouped and labeled.
6. Save your web to share with a classmate.

Modified Version A: Same as above except categories are given.

Modified Version B: Same as above but digital.

Exhibit 6.2 • Vocabulary Workstation—Memory Match Game

> **Memory match vocabulary game**
>
> Type of Workstation: Partner, group, or individual
>
> Materials: One deck of math vocabulary cards and a picture of the term
>
> Directions:
> 1. Spread all the vocabulary cards face down on the floor, desk, or table.
> 2. Take turns turning over two cards and trying to find a match.
> 3. When matches are made, students keep the cards.
> 4. The student with the most cards at the end wins.
>
> Modified Version A: Same as above except all cards are facing up the entire time.
>
> Modified Version B: Switch from partner game to independent station.

Individual Reference Folders

Individual reference folders* are a phenomenal tool to increase accessibility and personalize support for students discreetly. We love that students always have them and use them without prompting. You may be wondering, what is an individual reference folder, how is it managed, and how do students use it? Here is a detailed overview so you can experience their power for yourself.

Although we recommend a tangible folder, this may be digital for students who cannot access this format.

> **What?** Start with a two-pocket folder (one per student) that has brass fasteners to hold plastic page protectors. Skinny binders work but are bulkier. The number of sheet protectors varies, and you fill them as you move through your units of study. More on this later.

Why? The reference folder is intended to be a go-to tool for students to get themselves unstuck. It is relatively inexpensive to make, simple to maintain, and easy to personalize. It allows resources to be added or removed as needs change, and it is a powerful way to foster and expect independence during Math Workshop.

How? Once ready, fill the pages with whatever students need. Content varies by grade but think about templates you may have on hand, such as place value charts, multiplication charts, number bonds, number lines, fraction bars, coordinate grids, and other templates along those lines. The first page is intentionally blank for use as a dry-erase workspace that is always handy! The examples support math language and symbols. For additional examples, refer to the bonus resources.

When? As you establish routines with students (typically at the beginning of the year, but it can happen any time), demonstrate the role of the folder during Math Workshop. Share the expectation of always having the folder. This means they use it for workstations, Guided Math time, when playing games, or working on the computer. Stress that the folder is theirs, it is a tool, everyone has one, and they can use it any time. Normalizing tools as a universal practice, not a deficit-based one, is a critical element of Math Workshop Plus.

You may be thinking it would nice be to consolidate the many charts, templates, and reference sheets you have. Before you run out and start making fresh copies, consider how the structure of our sample pages fosters interaction and provides embedded support. As seen in Figures 6.8 through 6.13 in Table 6.1, except for explicit vocabulary pages, each page is intentionally designed to be *interactive*. Students access examples, definitions, visuals, and models, but most importantly, there is space for them to write and solve, often alongside a worked example. This is where the plastic sleeves come in. Students use these interactive spaces as dry erase boards and never need additional copies. If you don't love the idea of their work vanishing, an easy solution is to have students include a problem they found tricky on their accountability sheet, notebook, or exit ticket. This allows them to use the folder while maintaining accountability and producing work you can use formatively.

Table 6.1 • Math Reference Folder Examples

	Language Supports	Structural Supports
Primary	**Position Vocabulary** Figure 6.8 • Position Vocabulary	**Number Bond** Figure 6.9 • Number Bond
Upper Elementary	**Multiplication & Division** Figure 6.10 • Multiplication and Division	**Multiplication Chart - 12 × 12** Figure 6.11 • 12 × 12 Multiplication Chart

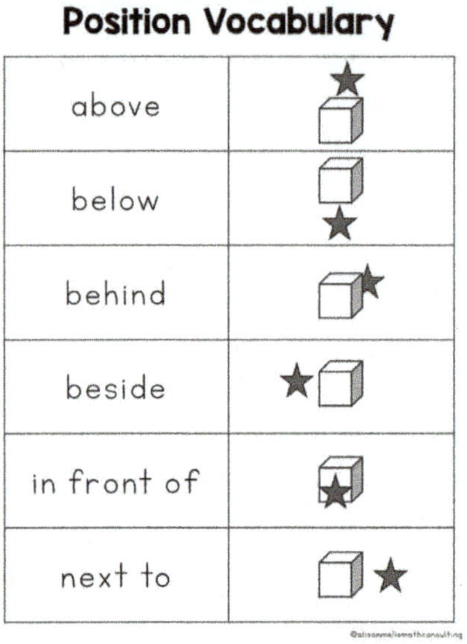

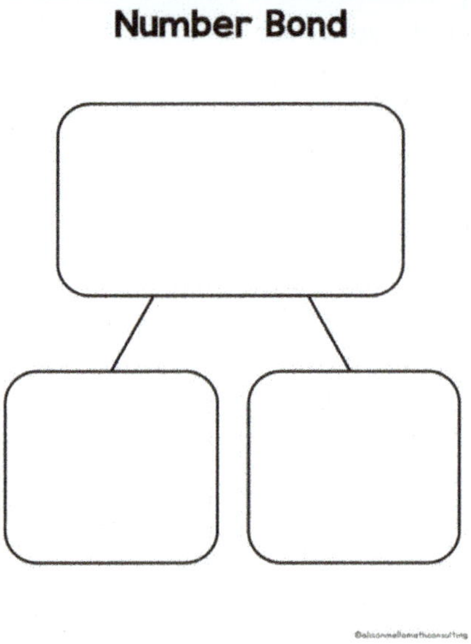

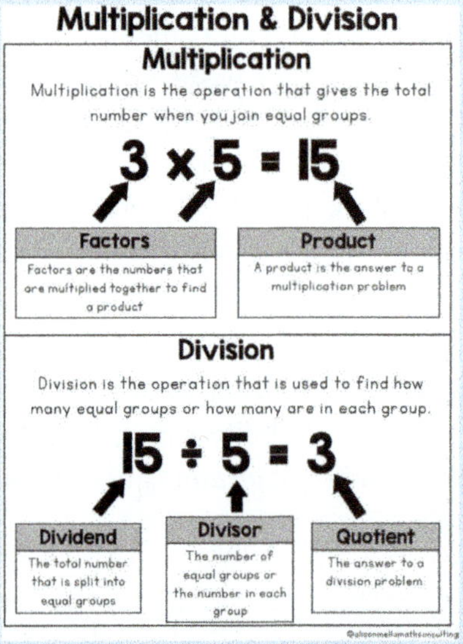

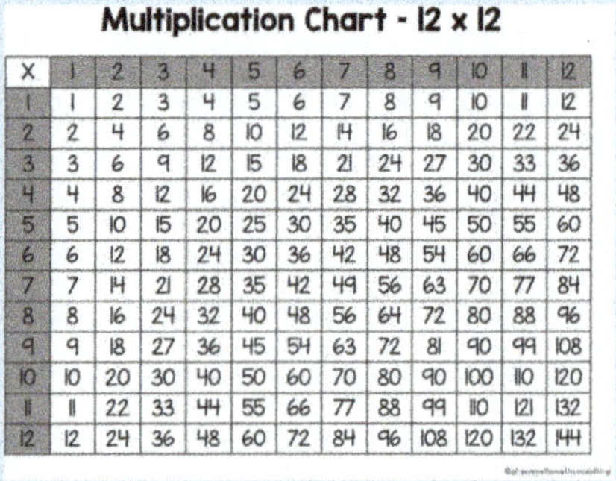

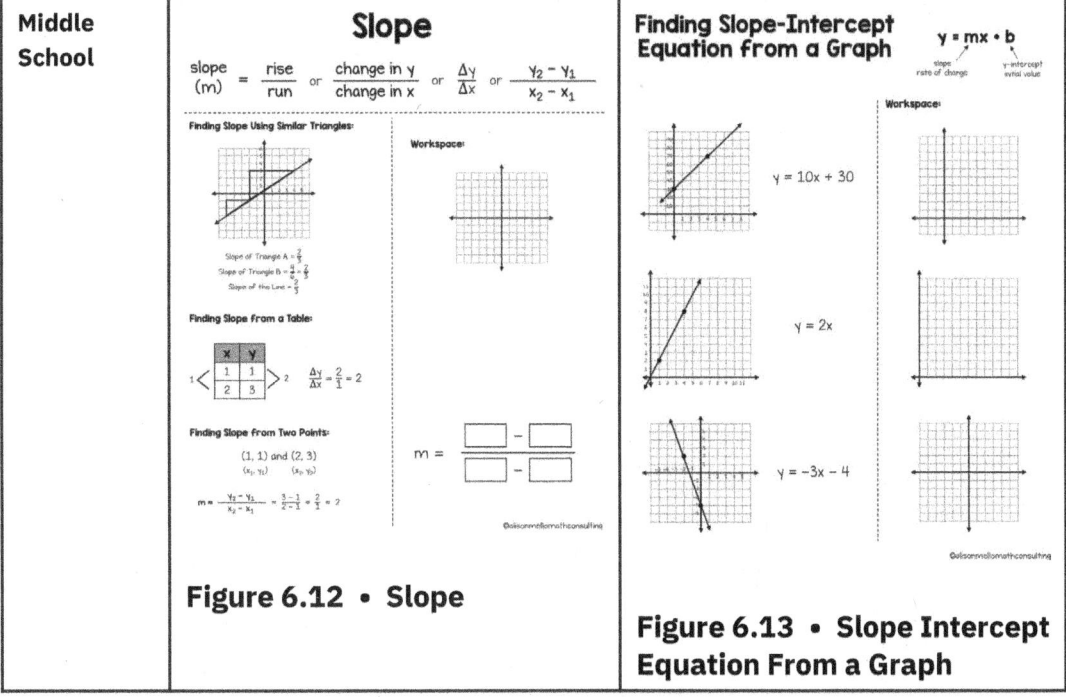

Figure 6.12 • Slope

Figure 6.13 • Slope Intercept Equation From a Graph

Here are some additional strategies to personalize reference folders for individual students:

- Add supports for prerequisite skills
- Add Latin roots
- Utilize color-coding
- Include translations next to English words where appropriate
- Include visual models for basic facts ($+ - \times \div$)
- Include visual representations of fractions, square numbers, measuring tools, etc.
- Include examples of standard form, expanded notation, scientific notation, etc.

> **Equity Check**
>
> Math Reference Folders are a wonderful tool to ensure all students have the personalized support needed to be successful in a way that is discreet. It's important to remember that equity and equality are two different things. Not everyone needs the same things to be successful, and providing what students need can, and should, be different.

Highlight Familiar Roots and Prefixes

Another way to support language comprehension is to highlight familiar roots and prefixes. In mathematics, prefixes often provide clues to indicate quantity. In geometry, the prefix may indicate the number of sides and angles in a polygon: tri-angle, quad-rilateral, octa-gon. In the metric system, the prefix indicates the unit of measure: centi-meter (100), kilo-gram (1000). Roots may give key information, such as with percentages—per-cent, meaning out of a hundred, and poly-nomial, meaning "many" in expressions with several different terms. Taking time to make these clues visible can accelerate comprehension and support decoding of unfamiliar words. Many teachers preview vocabulary in small Guided Math groups prior to a unit. This offers proximity to students so features of words can be explicitly emphasized and comprehension monitored more closely. See Figures 6.14, 6.15, and 6.16 for examples.

Figure 6.14 • Highlighting Roots and Prefixes in Triangle

Figure 6.15 • Highlighting Roots and Prefixes in Percent

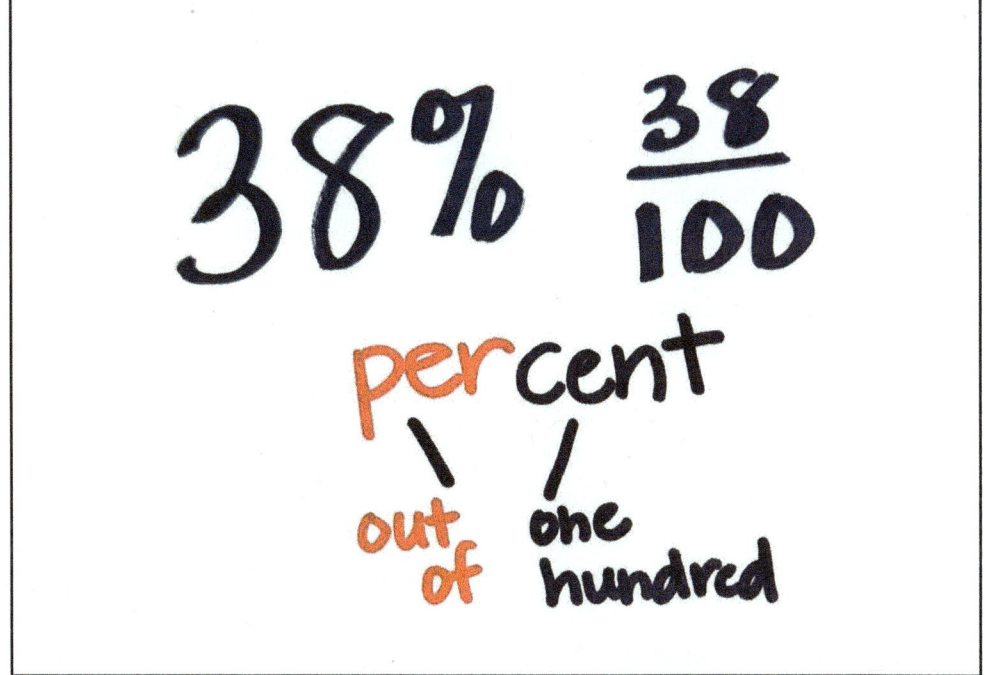

Figure 6.16 • Highlighting Roots and Prefixes in Polygon

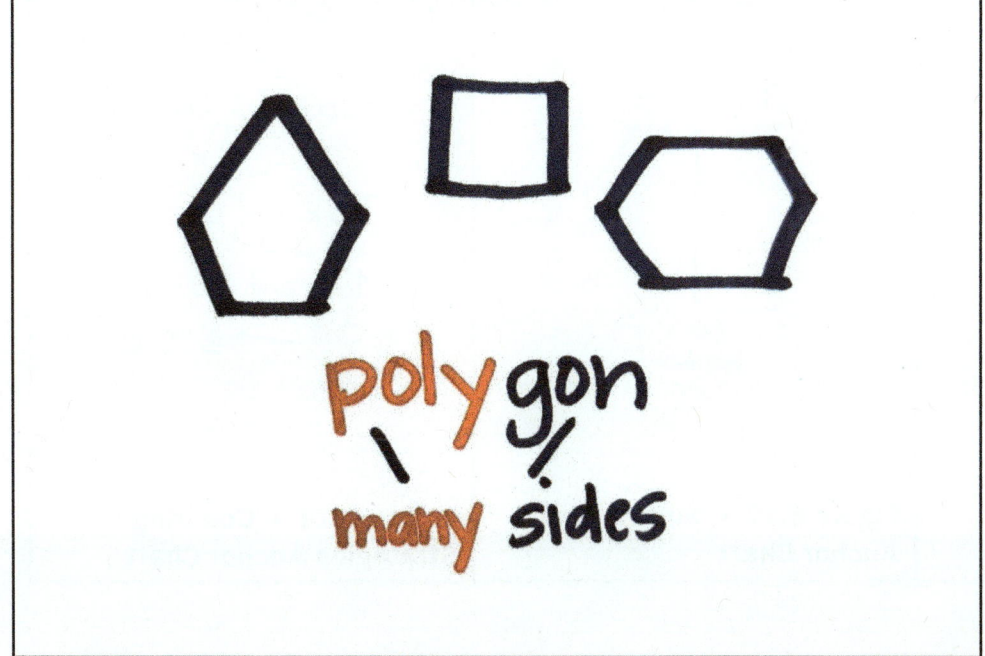

Chapter 6 • Design Options for Language & Symbols   **155**

Visual Anchor Charts

Anchor charts are a staple of any Math Workshop environment. During the First 20 Days, teachers generate norms and expectations with students and document common agreements on anchor charts. These may include guidelines for noise, getting started, remaining on task, using math talk, starting games, being a good partner . . . you get the idea. Other charts might outline success criteria, how to use math tools, or support math concepts and strategies. Anchor charts convey critical information and are an essential element of a successful Math Workshop, but have you ever thought about who *can't* access them? For example, if students are emerging readers or acquiring the English language, are these charts useful tools for *them*?

Think of your classroom. Do you use anchor charts? If not, we highly recommend that you add them. If you have them, what do they look like? Do they include visuals, or are they mostly words? Do they resemble the charts in Set A (Figures 6.17, 6.19, and 6.21) or Set B (Figures 6.18, 6.20, and 6.22) in Table 6.2?

Table 6.2 • Comparison of Anchor Charts to Support Vocabulary and Symbols

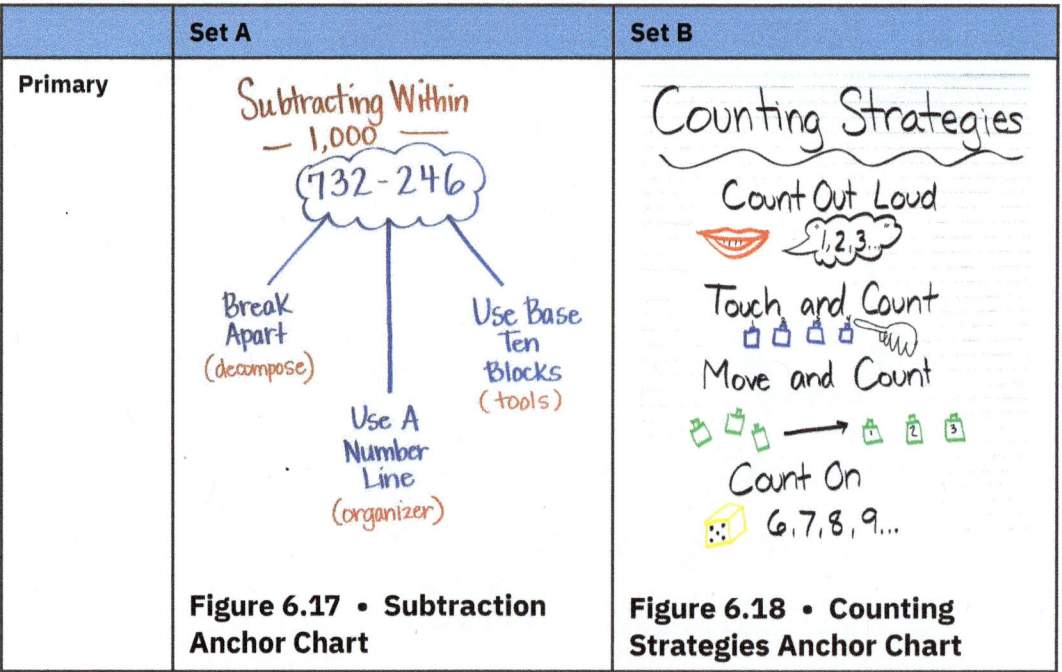

	Set A	Set B
Primary	Figure 6.17 • Subtraction Anchor Chart	Figure 6.18 • Counting Strategies Anchor Chart

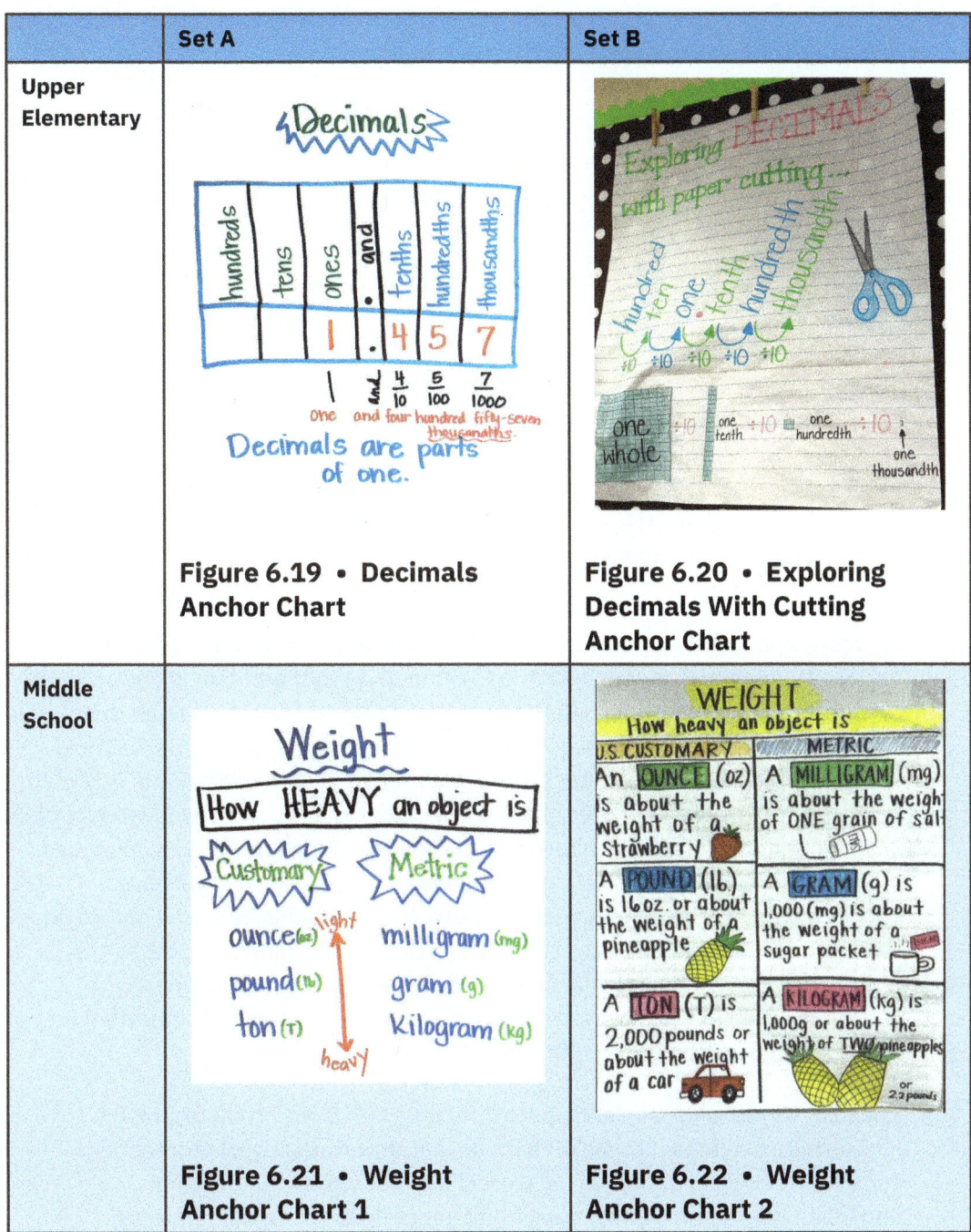

	Set A	Set B
Upper Elementary	Figure 6.19 • Decimals Anchor Chart	Figure 6.20 • Exploring Decimals With Cutting Anchor Chart
Middle School	Figure 6.21 • Weight Anchor Chart 1	Figure 6.22 • Weight Anchor Chart 2

Chapter 6 • Design Options for Language & Symbols

>
> ### Equity Check
> Access May Be Impacted by Language
>
> Language affects how students experience Math Workshop. From the launch to the mini lesson to workstations to Guided Math groups and the debrief, language is necessary for students to engage, communicate, and participate. Students for whom language is a barrier may not be able to participate without proper supports. Being aware of who needs visual models, hand signals, visuals, and translation tools can help students feel productive, involved, and successful. When supports are always present and used by *all* students, it creates an inclusive culture that celebrates and embraces variability.

Support Decoding of Text, Mathematical Notation, and Symbols

In previous sections, we discussed strategies to make mathematical notation, symbols, and words more accessible. What we haven't explored yet is what happens when these elements come together in a word problem. Many students dread word problems, especially when they pop up on tests or independent practice. The complexity of decoding and comprehending the words, coupled with relating that to mathematics, presents multiple potential barriers. As noted by Braselton and Decker (1994), reading in math class "is a complex mixture of words, numbers, letters, symbols, and sometimes graphics" (p. 276). Decades of research exploring the relationship between language and math have concluded that early numeracy, particularly counting skills, is a predictor of later reading skills (Leppanen et al., 2006); performance on math word problems is strongly related to performance in reading comprehension (Vilenius-Tuohimaa et al., 2008); and knowledge of vocabulary is critically important in understanding math (Boulet, 2007).

Because teachers usually anticipate difficulty with word problems, many share strategies meant to make them easier. Unfortunately, well-intended shortcuts, such as finding "signal words" or "key words," undermine and limit understanding and may inhibit sense-making. In the following problem, the signal word is *all*, which students are taught to interpret as addition.

> **Bo sorted 30 candies into boxes. If each box held 5 candies, how many boxes did he have in all?**

Is the answer 35 boxes? Would that make sense? Do you think students would they see "all" and mindlessly add? Understanding what the language means *in the context of the story* is required for sense-making. When used in isolation, methods like C.U.B.E.S (circle, underline, box, eliminate, solve) eliminate thinking and make word problems feel like a technical exercise, rather than a reasoning one (Goldberg, 2003).

We must recognize when our practices don't move students forward. For example, how many of us read the problems for students? While this may be appropriate in kindergarten and part of first grade, to become proficient, students must have the opportunity to do it themselves. We suggest two mathematical language routines designed to amplify language and help students tackle word problems through comprehension and reasoning. The first is Numberless Word Problems, which we learned about from Brian Bushart (numberlesswp.com), and the second is the 3 Reads Protocol (Kelemanik et al., 2016). See Table 6.3 for a description of each routine and consider how it might look in your context.

Table 6.3 • Mathematical Language Routines

	Numberless Word Problems	**3 Reads Protocol**
What?	A word problem without numbers	A protocol for word problems that includes reading the math problem 3 times, each time with a different goal.
Why?	Without numbers to "pluck and plug," students reason about the story. They make sense of the language to understand what is happening without rushing to calculate.	This protocol has students read the problem 3 times before they solve.
How?	1. Remove the numbers from a word problem. 2. Example: 3. *Original*: Jen had 5 lollipops and 3 candy bars. How many more lollipops than candy bars did she have?	1. Students read the problem to understand the context and what is happening. *When introducing the strategy, it is helpful to remove the question as in the numberless word problem. 2. The first read can be followed by discussion to clarify what is happening in the story.

(Continued)

(Continued)

	Numberless Word Problems	3 Reads Protocol
	4. *Numberless*: Jen had some lollipops and some candy bars. She had more lollipops than candy bars. 5. Use questioning to have students share what they know, what they wonder, and what they think the question is. A great discussion should ensue. 6. Give small bits of additional information until the entire problem/question is revealed. 7. Have students work together or independently to solve.	3. Students read the problem a second time (again, without the question) to focus on the quantities, units, and other math concepts. 4. Use questioning to draw out the ideas. 5. The third read is where students consider the question being asked and formulate a plan/strategy to solve. 6. The process is complete when students solve the problem and check their solution.

Cultivate Understanding and Respect Across Languages and Dialects

When you think about how much language students encounter in math, it's daunting! While some words have clues as to their meaning, some are completely unfamiliar. In these instances, we may have to get creative about how we support comprehension. To think about connecting words and phrases across languages, let's explore some ideas and strategies you can integrate immediately if you haven't already.

Use Cognates

The population of multilingual learners (ML) in the United States continues to grow. The percentage of ML students grew from 8.1% (or 3.8 million students) in 2000 to 9.6% (or 4.9 million students) in 2016, and it is projected to reach 25% of the student population in 2025 (McFarland et al, 2019). With increasing diversity, we are challenged to integrate new strategies to ensure access for all students. For those acquiring the English language, one evidence-based strategy is the use of cognates. Cognates, specifically "true" or "friendly" cognates, are words that are spelled and pronounced similarly in two languages and have a similar meaning in both (Gómez, 2010). Cognates create access when students recognize the words and meanings, even if they do not yet fluently speak English. If cognates are new to you, refer to Table 6.4 to see the examples provided by Gómez.

Table 6.4 • Examples of English- and Spanish-Friendly Cognates

English	Spanish
division	división
hexagon	hexàgono
congruent	congruente
circle	circulo
prism	prisma
convex	convexo

Use Comprehensible Input

If you aren't familiar with comprehensible input, it's not as complicated as it sounds. Simply stated, comprehensible input is an instructional technique that helps students interpret language even if they can't yet fully produce it (Krashen,1982). What does that look like in a math classroom? At the risk of oversimplifying, highly effective methods of comprehensible input include using gestures and visuals, employing hands-on materials, integrating multisensory methods, using repetition, and clarifying symbols. You may be feeling relieved that these are things we have been discussing and perhaps you already use.

Since we've already discussed hands-on materials, visuals, and multisensory methods, let's take some time to explore using gestures, repetition, and clarifying symbols. Let's begin with gestures.

Use Gestures

As noted by Wakefield et al., "[C]hildren who learned the math task while gesturing seemed to incorporate that gesture into their lasting understanding of how to solve problems." Incorporating gestures, such as those in Figures 6.23 to 6.31, engages the motor system and neural systems, and gestures can "facilitate learning even in the absence of a spoken strategy" (2019, pp. 2350–2351).

Examples of Gestures for Operations

Figure 6.23 • Addition

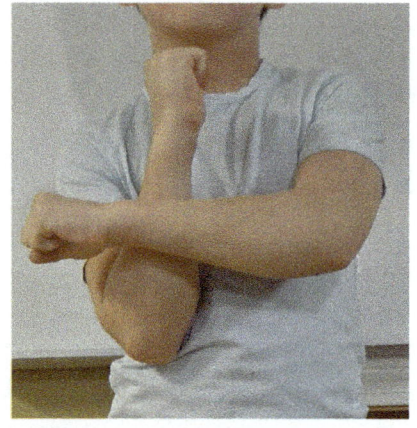

Figure 6.24 • Subtraction

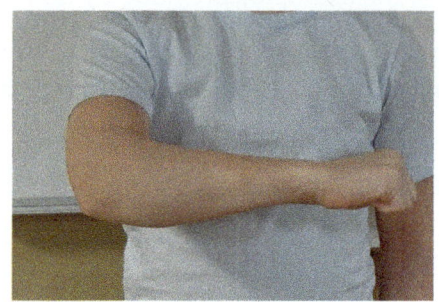

Figure 6.25 • Equal To

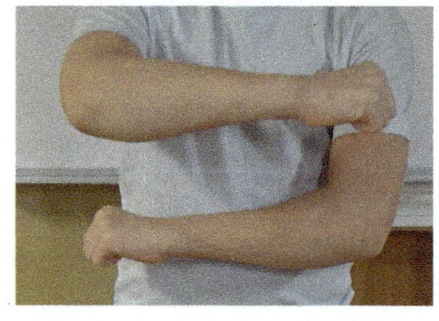

Figure 6.26 • Acute Angle

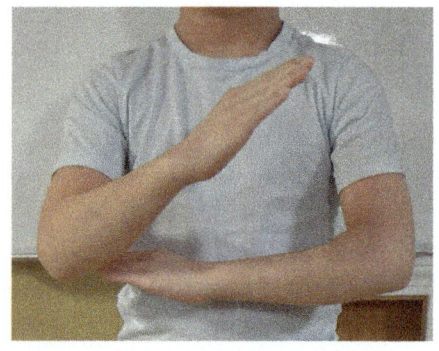

Figure 6.27 • Obtuse Angle

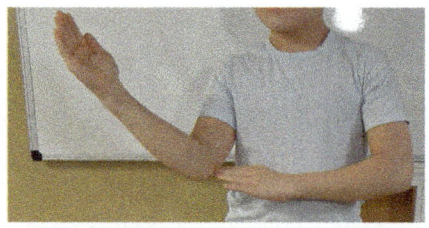

Figure 6.28 • Right Angle

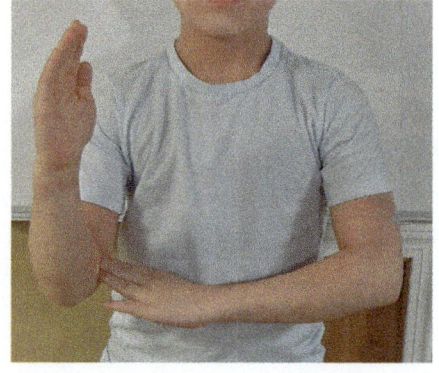

Figure 6.29 • Parallel Lines

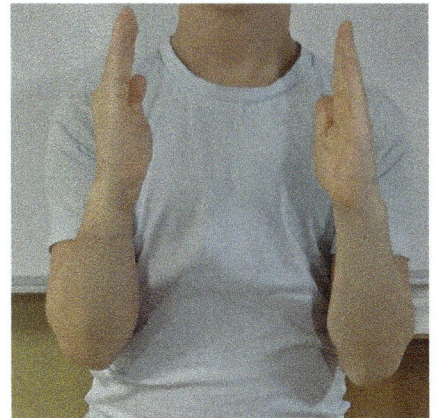

Figure 6.30 • Negative Slope

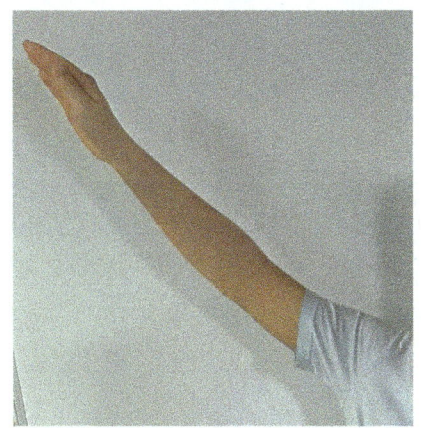

Figure 6.31 • Positive Slope

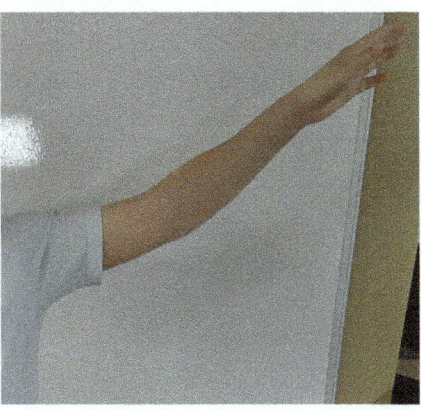

Use Repetition

This strategy is likely alive and well in your classroom, but Table 6.5 has some examples in case you need them. In Math Workshop, repetition happens in workstations that revisit essential skills and standards, or routines that are in regular rotation. Since we are thinking about language here, workstations that revisit vocabulary could include matching games, scavenger hunts, and word sorts. Those are great, so if this is new to you, start there. Table 6.5 offers some strategies that don't require prep and should fit well with what you already do.

Table 6.5 • Strategies for Repetition to Support Mathematical Language, Expressions, and Symbols

Strategy	Instructions
Point	Each time a vocabulary word comes up, point (either with your finger or a pointer) to the word on your word wall
Call and response	Say the word—students repeat
	Say the definition—students repeat
Songs or rhymes	Share songs or rhymes that may be repeated to strengthen retention of the meanings of words

Clarifying Differences Between Symbols

Let's view the progression of multiplication symbols through the eyes of our students. In second grade, multiplication is introduced as repeated addition using the plus sign. In third grade, you turn that plus sign 45 degrees to the right, and it becomes a multiplication sign. When using that sign, the addend becomes a factor, and it's not repeated. It looks like the letter x, but it means something different. If you wait a few years, the x turns into a dot. And in seventh grade, it disappears completely when a number is next to an open parenthesis, or if a number is with a variable, because multiplication is implied. Oh, and within that time frame, the x comes back, but it no longer signals multiplication because it's a variable. What? Wow, it's confusing! There are lots of symbols in math, and we know you're shocked that they don't all make sense.

Division also shows up in different ways. Each of the following expressions represents 1 divided by 4:

$$1 \div 4$$

$$\frac{1}{4}$$

$$4\overline{)1}$$

How many students don't interpret a fraction as division? How does that affect operations with complex fractions in Grade 7? Imagine seeing the problem that follows and having no idea that it represents dividing a fraction by a fraction.

$$\frac{\frac{1}{2}}{\frac{3}{4}}$$

Adding a page to the Reference Folder to support symbols is a great way to build comprehension and help students navigate problems independently. Make a list of symbols used in your grade(s) and attach meaning by grouping them and attaching visuals. Symbols such as $<, >, =, +, -, \times, \div, \pi, \approx, \sqrt{}, \Sigma, \leq, \geq, \%$, and $ can look completely unfamiliar to students. Other symbols may be more subtle because they look familiar but mean something different. Examples include exponents, cube roots, and if you teach high school, exclamation points.

Address Biases in the Use of Language and Symbols

We're sure you can see the intentionality with which we believe language should be approached in a Math Workshop Plus classroom. Language is a place where biases can pop up, so taking the time to reflect on how to address inherent biases in the language and symbols used in your resources is an important move in barrier removal. Such biases can mask what students know and understand. Have you thought about this before? If so, how have you approached these issues? If not, explore the following ideas and see what could work best in your environment:

- Encourage students to add labels to vocabulary in their native languages.
- Capitalize on opportunities for students to share symbols and language for math items (such as currency or numerals) from their cultural context.
- Use closed captioning in multiple languages.
- Allow students to label classroom objects and areas with the words from their language.

These intentional actions honor all students and signal that the identities of all students are valued. As the classroom becomes more language-rich, accessibility increases for students, and the classroom community grows stronger. Addressing biases in the use of language and symbols fosters a sense of belonging for all members of the learning community and diminishes the perception that other languages are inferior.

Illustrate Through Multiple Media

One benefit of teaching in the 21st century is the plethora of digital tools available to enhance instruction. While we love physical manipulatives, paper folding, graphing calculators, and graphic organizers, digital tools are versatile, free, easy to manage, and can remove a variety of barriers. These applications decrease the tendency to deliver content primarily through text and provide "no-prep" opportunities to infuse visuals, videos, simulations, and virtual manipulatives. These representations support comprehension, connections, and allow students to pursue and demonstrate proficiency in different ways.

When concepts are developed using multiple media, we can talk less and get students thinking. For example, look at Figures 6.32 through 6.40 within Exhibits 6.3, 6.4, and 6.5. If you put any of those images up and simply asked students what they notice and wonder, what responses would you anticipate?

Exhibit 6.3 • Splat Math

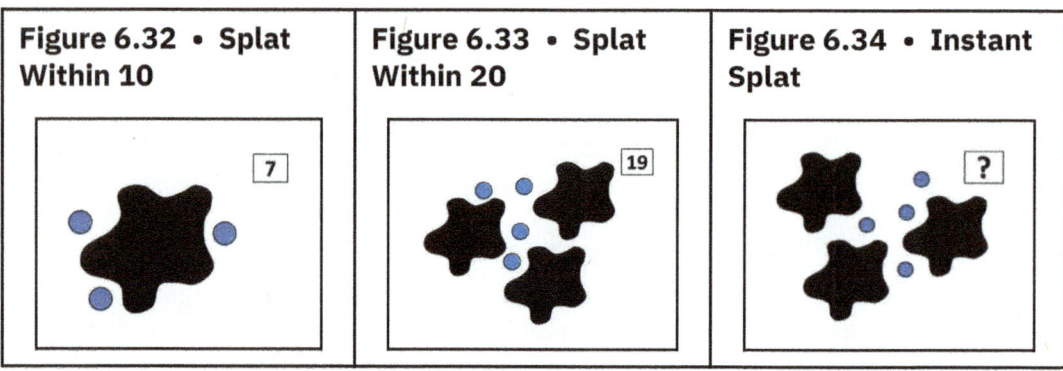

Credit: *Steve Wyborney Splat!*

Exhibit 6.4 • Digital Number Balance

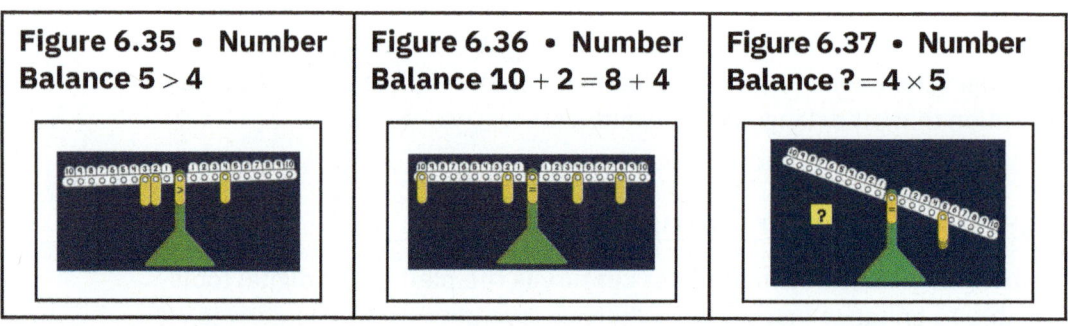

Credit: Didax Virtual Manipulatives

Exhibit 6.5 • Digital Fraction Bars

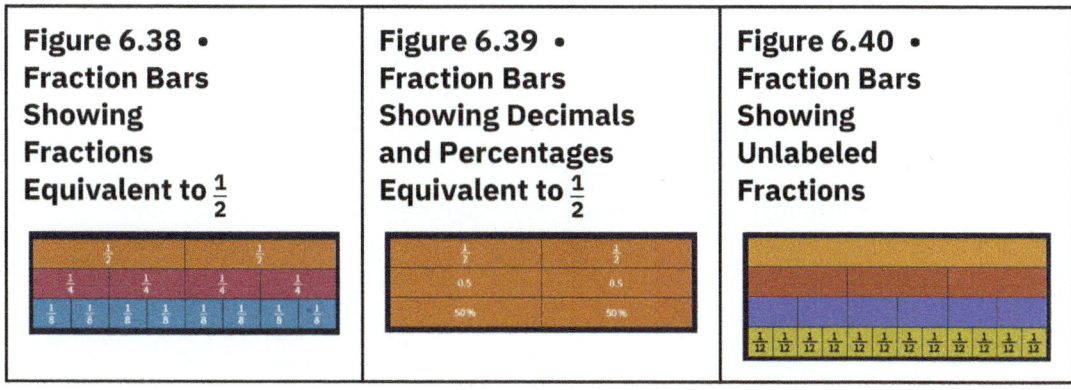

Figure 6.38 • Fraction Bars Showing Fractions Equivalent to $\frac{1}{2}$

Figure 6.39 • Fraction Bars Showing Decimals and Percentages Equivalent to $\frac{1}{2}$

Figure 6.40 • Fraction Bars Showing Unlabeled Fractions

Credit: *Mathigon Polypad*

What did *you* notice? What did *you* wonder? How much math is communicated in the previous exhibits without any words? If you are a primary teacher, what might your students notice? Do you think you'd hear math vocabulary such as *greater than, less than, add, subtract, or equal to*? Even though they have not studied fractions, multiplication, or percentages, do you think they would notice some relationships? Do you think students could work backward from the Splat or presume since there are four identical splats they all must be the same amount? Could they reason through that? Wouldn't they, and you, be amazed to solve $4x + 7 = 19$? What about the Number Balances? Could students use skip counting to conclude that the right side equals 4×5, or 20, and therefore any combination of numbers that add to 20 on the other side makes a true statement? What might they notice about the fraction bars and how the 2, 4, and 8 are related?

Illustrating concepts using a variety of media allows students to explore without language getting in the way. Digital tools don't make noise or fall on the floor, images may be made larger or smaller, and objects can be manipulated to highlight connections between and among concepts. These tools also allow students to make connections to what they know and expand upon it, rather than being shown a procedure to get an answer.

For visually impaired students, consider what other tools they can engage with to benefit from this exercise. Many digital resources have accessibility features such as screen readers, contrast controls, and screen magnifiers.

Say Bye-Bye to Barriers When You . . .

- Post language frames and sentence stems
- Make vocabulary visual
- Create individual reference folders
- Highlight roots, prefixes, and cognates
- Create visual anchor charts
- Incorporate gestures
- Leverage multimedia to reduce language during instruction

Action Plan

Assess your current supports for Designing Options for Language & Symbols. Download the Chapter 6 Self-Assessment at https://companion.corwin.com/courses/MathWorkshopPlus.

Try It! Supporting Mathematical Language and Symbols

Select and try three new strategies to support mathematical language and symbols. Monitor for differences in participation, output, and interaction.

Connections

Share something you're excited about with a colleague or with the community via social media using the #mathworkshopPLUS.

◇◇◇◇◇◇◇◇◇◇◇◇◇◇◇◇◇◇◇
Chapter 7

Design Options for Building Knowledge

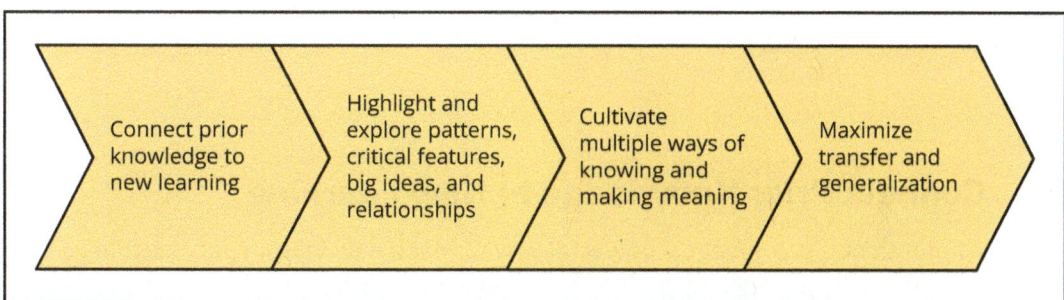

The call of the 21st century is to have the ability to transfer learning to new and different contexts. To achieve this, students need frequent opportunities to acquire and apply knowledge. Information processing skills are built over time through experiences and activities that help students to transfer knowledge. Math Workshop Plus offers students plenty of opportunities to learn how to focus, to make connections between what they know and what they are learning, to categorize information, and to build on that information to do new and different things. Your teacher moves can either promote or inhibit the ability of students to do this.

As you think about *Design Options to Build Knowledge* for all students, take time to pause and reflect on the following questions and the teacher moves that support these actions:

1. *Connect prior knowledge to new learning*: How much do I know about what math students have previously learned so I can activate it and support connection?

2. *Highlight and explore patterns, critical features, big ideas, and relationships*: Is there an ongoing effort to seek and find patterns—to show and make connections between math big ideas and mathematical relationships? Do I feel like I know the patterns, connections, and relationships?

3. *Cultivate multiple ways of knowing and making meaning*: Do I offer different entry points to lessons and vary the way I present information?

4. *Maximize transfer and generalization:* How do I position students to apply knowledge and generalize ideas in unfamiliar contexts?

Connect Prior Knowledge to New Learning

As mentioned previously, a powerful way to make new learning accessible to students is by linking it to prior knowledge. Math Workshop Plus offers multiple opportunities for you to do this. For example, you can use the launch to make these connections (CAST, 2024; Marzano, 2001); prior knowledge prompt cards, prior knowledge mind maps, and photos paired with curious questions are great ways to spark prior knowledge (Newton, 2023). Making explicit interdisciplinary connections by using poems, songs, and picture books opens up the conversation and supports transfer. Providing multiple entry points to a lesson and optional pathways through content invites all learners to the table. Things like introducing and exploring content through art, film, and literature (CAST, 2024) create on-ramps for learners who may not see themselves through a mathematical lens (yet!). See Table 7.1 to explore some ideas to activate, and support connections to, the prior knowledge that your students bring to your class.

Table 7.1 • Prior Knowledge

Part of the Lesson	Strategy	Instructions
During the lesson	Prior Knowledge Picture	Give students a blank piece of paper and have them write everything that they know about the topic.
During the launch and workstations	What do you know? Brainstorm	Students work with a partner and brainstorm with words and pictures everything they know about the topic.
During the launch and the debrief	Rate yourself!	Ask your students: How much do you know about _____? Have them vote with their fingers: 1, 2, 3 . . .
During the launch	Anticipation Guide	Make a guide with true/false or yes/no questions and have the students respond with their thoughts; then these are discussed.
During the launch, workstations, or the debrief	Mindmap of What You Know	Give the students a mindmap (map with a thinking person), and they have to put everything they are thinking about the topic on the map.
Throughout the lesson at different times to track knowledge growth	Quiz (Give at different times throughout the unit during the launch or debrief to track the knowledge growth)	Give students a quick quiz to find out what they know and what they are wrestling with.
During the launch	KWHL (know–want to know–how do I find out–learn)	Do a KWHL chart with the class and leave it up during the unit of study and add to it as they learn things.
During the student activity	Quick conference or 5-minute interview	Meet with students briefly and ask them what they know.
Up and posted; often completed during workstations; they could also add to it after a Guided Math lesson	Graffiti Wall	Have students go up to the wall and write and draw pictures of what they know about the topic. Leave the wall up throughout the unit and add to it as the class learns things.

 Download this Prior Knowledge Table at https://companion.corwin.com/courses/MathWorkshopPlus.

Exhibit 7.1 is an example of a Question Organizer for tapping into prior knowledge. When you share this with students, have them answer all of the questions and share their thinking with a neighbor before they come back to a whole-class discussion.

Exhibit 7.1 • Question Organizer—Tapping Into What I Know!

Tapping Into What I Know!		
Have you heard of this topic before?	What did you do with this topic last year in _____ grade?	What are some words you know about this topic?
What are some images that come to mind about this topic?	Where do we see this in our everyday life?	What does your partner know about this topic?

 Download this Question Organizer at https://companion.corwin.com/courses/ MathWorkshopPlus.

Use Visual Imagery

Your students live in a visual world. They look at phones, computer screens, tablets, books, videos, and television. Through these media, they explore their interests and make sense of the world around them. How often do you incorporate visual imagery to anchor thinking and support connections to prior knowledge? Capitalizing on what students are familiar with increases engagement, all while building connections and making new learning more accessible.

Figure 7.1 shows a counting example in which someone takes a doughnut out of a doughnut box. The minute this picture is shown, students are excited and ready to discuss it, and maybe you are too! Maybe you like doughnuts and maybe you don't, but when students recognize things like this from their everyday lives, they are primed to learn. If you teach kindergarten, would students notice how the 6 doughnuts match the arrangement of 6 dots on dice? Could they then be ready to tell a subtraction story that they understand, know about, and can instantly relate to? Jumping into a unit on subtraction this way levels the playing field. Everybody joins in because they see where they fit in. This picture springboards the class into a very rich discussion about subtraction.

Figure 7.1 • Counting Doughnuts Task

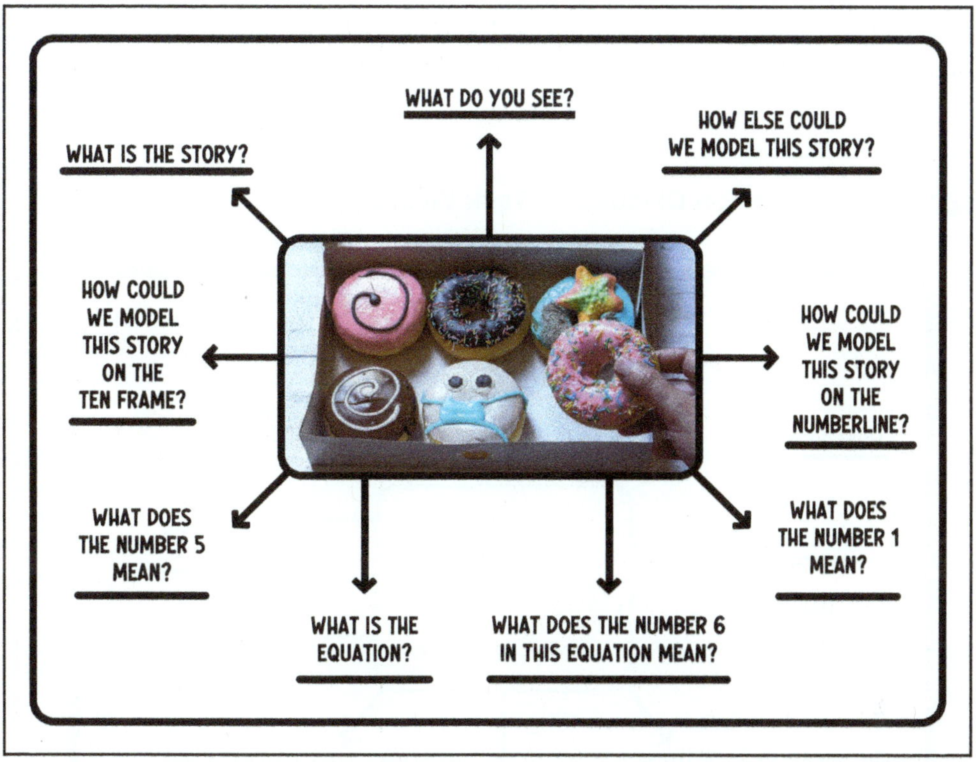

[Figure: A central photo of a box of six doughnuts with arrows radiating outward to eight prompts:
- WHAT DO YOU SEE?
- WHAT IS THE STORY?
- HOW ELSE COULD WE MODEL THIS STORY?
- HOW COULD WE MODEL THIS STORY ON THE TEN FRAME?
- HOW COULD WE MODEL THIS STORY ON THE NUMBERLINE?
- WHAT DOES THE NUMBER 5 MEAN?
- WHAT DOES THE NUMBER 1 MEAN?
- WHAT IS THE EQUATION?
- WHAT DOES THE NUMBER 6 IN THIS EQUATION MEAN?]

Source: Newton, 2024.

Visual images enrich the learning experience for many students and build connections not only in their brains but to each other. Activities such as this help students to find commonalities with their peers and can be an effective way to highlight cultural aspects of life. Infusing visuals can happen in the introduction, during problem-solving, in Guided Math groups, or any time!

Figure 7.2 offers an example of what this might look like with fractions. By taking a complex topic and connecting it to the everydayness of life, students make connections to how much they already know, even if they don't realize they know it! Think about the difference between writing $2\frac{1}{2}$ on the board and projecting an image of $2\frac{1}{2}$ apples. When you put apples on the board, it feels familiar, and students aren't intimidated. Everyone is in and can make connections to what they know. This is the essence of an asset-based lens. No one really knows more about these apples than anyone else. Everyone feels as if they can enter the conversation in a way without feeling or looking awkward

Chapter 7 • Design Options for Building Knowledge

because Jamal doesn't know any more about apples than Jenny. Starting from what students know sparks a rich discussion and allows you to build a bridge between prior knowledge and current knowledge with the full attention and engagement of your students. If you've ever taught fractions, you know that isn't typically the case!

Figure 7.2 • Fraction Discussion Task With Apples

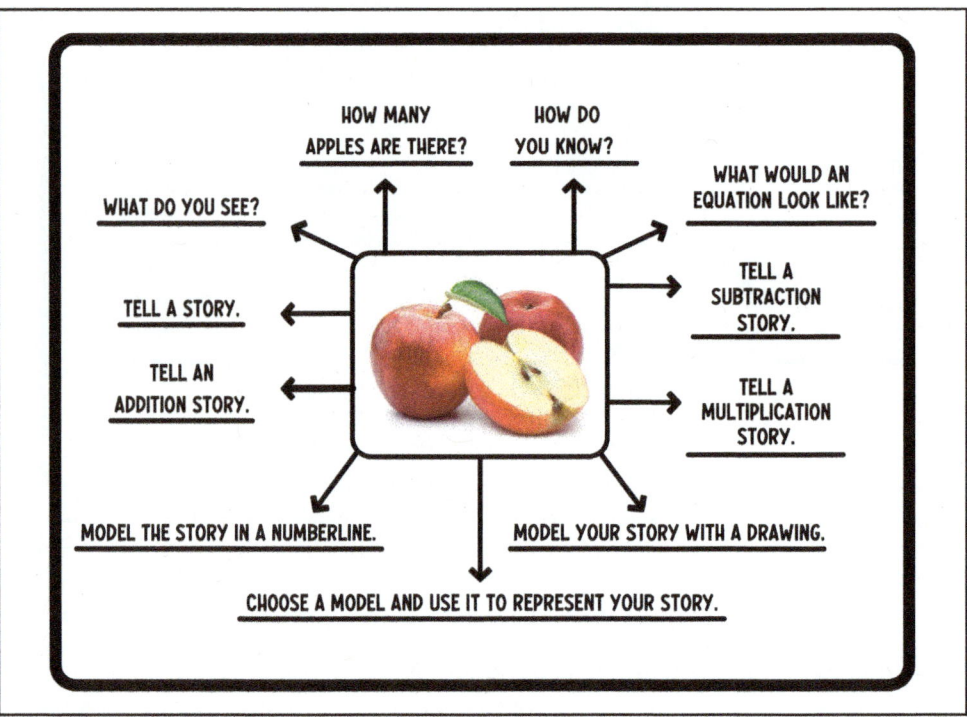

Newton, 2024

Use Advanced Organizers

Different types of advanced organizers help students to see information in ways that make sense. They are "a kind of cognitive bridge, which teachers use to help learners make a link between what they know and what is meant to be learnt" (Mostafa, 2017, para. 1). Graphic organizers visually help students to make connections between "facts, terms and ideas within the learning task" (para 1). They can be a great tool in workstations to capture the thinking of students. They also work well in small groups because you can listen to students as they explain how they have organized their thinking. See the following examples for primary, upper elementary, and middle school, respectively. In the primary example (Figure 7.3), students are presented information about subitizing, with various visual examples all connected within one organizer. In the upper elementary example (Figure 7.4), various models are connected around the big idea of multiplication. Students are making connections between the models, the vocabulary, and strategies. In the middle school example (Figure 7.5), students are diving deep into the concept of comparing decimals from the perspective of various models. Can you see how this supports connections between the hundredth grid, actual money, the number line, and the math notation? A graphic organizer is a way of saying, "Come take a walk with me, and let's talk about and make connections with what we can see."

Figure 7.3 • Advanced Organizer for Primary Grades on Subitizing

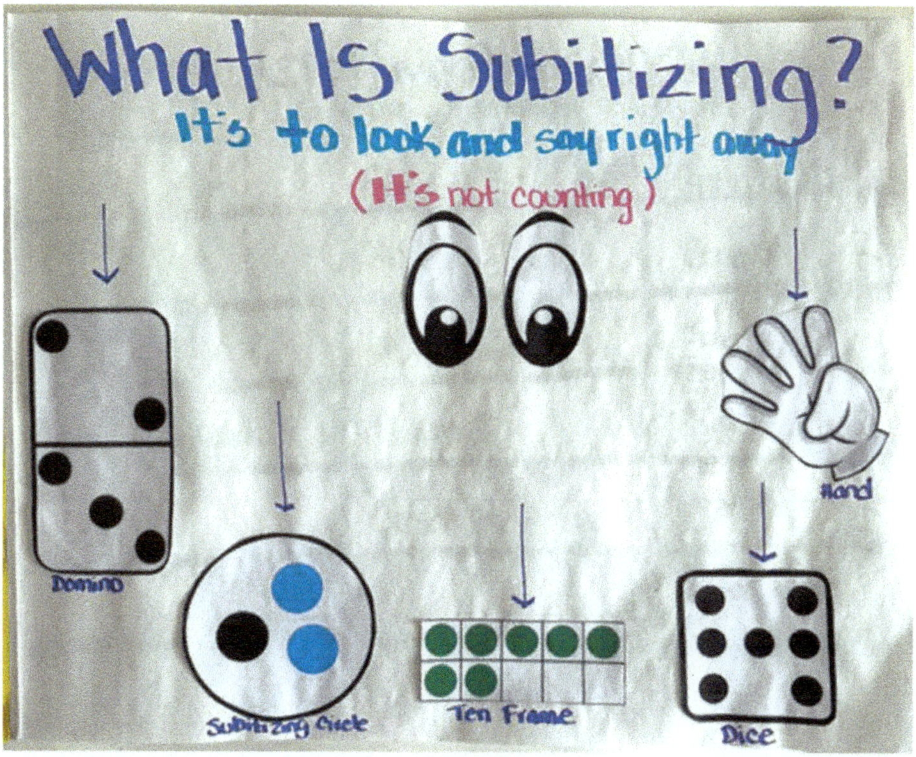

Figure 7.4 • Advanced Organizer for Upper Elementary Grades on Multiplication

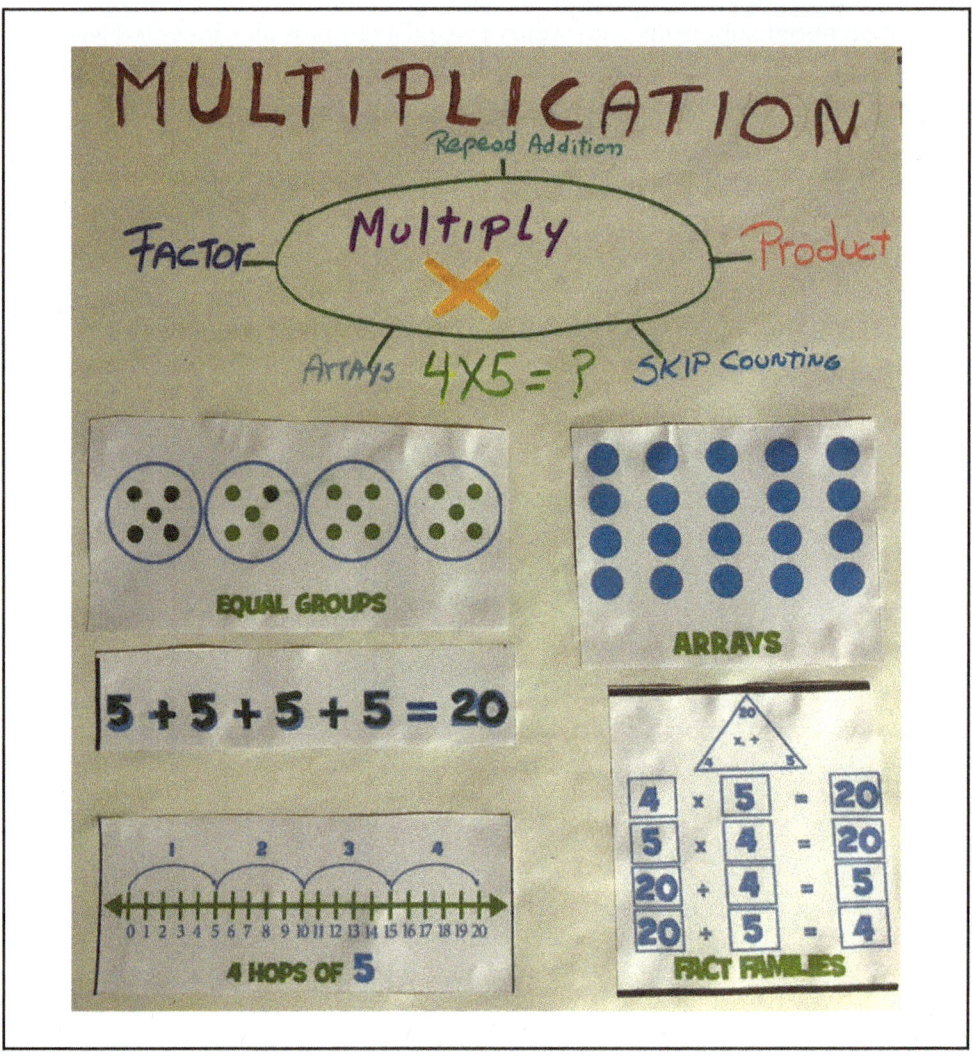

Figure 7.5 • Advanced Organizer for Middle School on Comparing Decimals

Comparing Decimals

HUNDREDTHS CHART
0.01–1.00

0.01	0.02	0.03	0.04	0.05	0.06	0.07	0.08	0.09	0.10
0.11	0.12	0.13	0.14	0.15	0.16	0.17	0.18	0.19	0.20
0.21	0.22	0.23	0.24	0.25	0.26	0.27	0.28	0.29	0.30
0.31	0.32	0.33	0.34	0.35	0.36	0.37	0.38	0.39	0.40
0.41	0.42	0.43	0.44	0.45	0.46	0.47	0.48	0.49	0.50
0.51	0.52	0.53	0.54	0.55	0.56	0.57	0.58	0.59	0.60
0.61	0.62	0.63	0.64	0.65	0.66	0.67	0.68	0.69	0.70
0.71	0.72	0.73	0.74	0.75	0.76	0.77	0.78	0.79	0.80
0.81	0.82	0.83	0.84	0.85	0.86	0.87	0.88	0.89	0.90
0.91	0.92	0.93	0.94	0.95	0.96	0.97	0.98	0.99	1.00

Money

Using Symbols

.38 < .55

Number Line

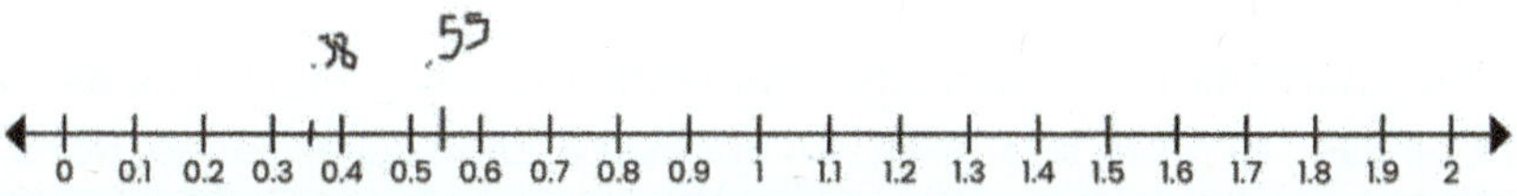

Source: Dr. Nicki Newton, 2024. Penny icon by Istock.com; dime icon by Istock.com/KavalenkavaVolha.

Connections to SEL

Prior Knowledge

When you tap into students' prior knowledge, you assist in building and promoting students' self-awareness. Asking students to think about what they know, not only from a book but from their lived experiences, honors them as holders of knowledge. Students bring a wealth of knowledge to school with them (Gonzalez et al., 2005). When you honor this knowledge, students learn to be proud and excited about the things that they *do* know, and the ways in which they can connect that knowledge to new.

Equity Check

Prior Knowledge

Every student comes into your classroom with prior knowledge. When you recognize this and use it to prepare students for new learning, it levels the playing field, and everybody can enter into the discussion with confidence. When you look at lessons through an equity lens, remind yourself to start by finding out what students already know and considering what lived experiences students bring that connect to the lesson.

Highlight and Explore Patterns, Critical Features, Big Ideas, and Relationships

Did you know that math has been referred to as the "science of patterns" (Resnik, 1981). Students who identify patterns are more likely to recognize connections in mathematical representations and develop strong algebraic reasoning skills (Smith et al., 2007). Patterns help students to use their prior knowledge to make sense of new ideas. In your practice, how do you highlight patterns, relationships, and big ideas to help students successfully navigate and make meaning of what they are learning?

If you create charts, tables, and posters for Math Workshop Plus, consider how these can be used to prompt students to uncover and discover connections, capture big ideas, and organize important information. These resources can then be used as references when students are working in math workstations. Structures like these help students determine what information is important and use examples and nonexamples to assimilate new information.

Use Graphic Organizers

You probably use strategies to help students zoom in on important information. As you present "text, graphics, diagrams, and formulas" throughout the mini lesson, Guided Math group, workstations, and the debrief, you might highlight, underline, embolden, star, circle, and accentuate the essential information in a variety of ways (CAST, 2024). Examples of this are included in Figure 7.6 and 7.7. Depending on the grade you teach, you may do this with students early in the year so that they develop the capacity to do it themselves. In the primary grades' example of a graphic organizer, students are given the opportunity to practice subtraction models in a variety of ways, including with ten frames, a math sketch, a number line, a number sentence, counters, and number bonds.

Figure 7.6 • Primary Example of a Graphic Organizer for Modeling Subtraction

I Can Model Subtraction

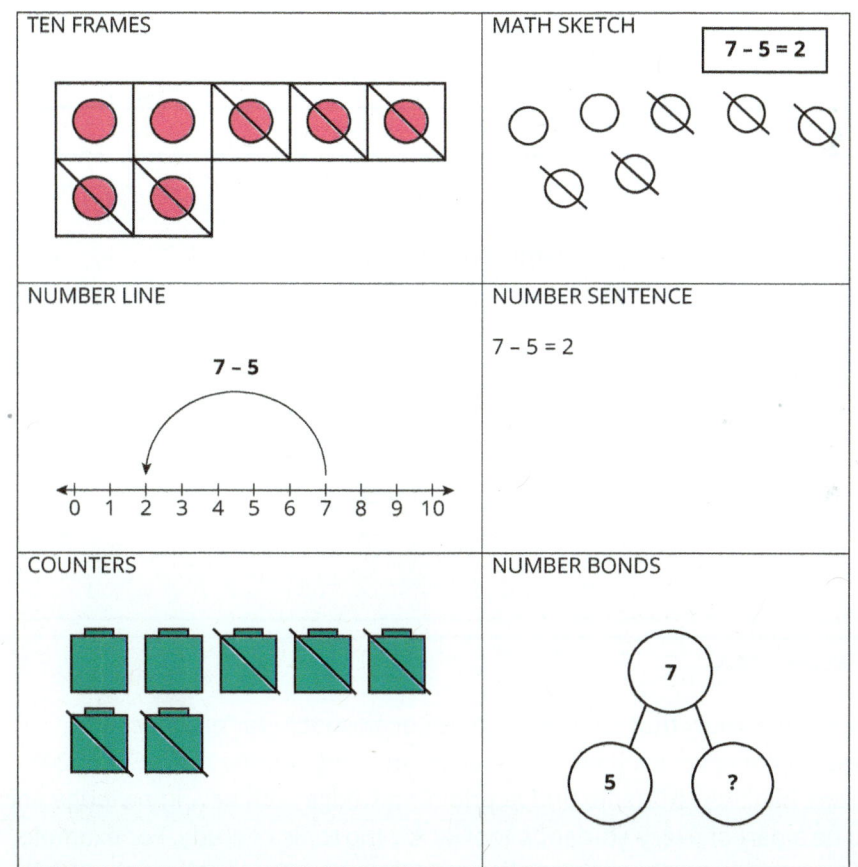

Source: Newton, 2024.

Tape diagrams are a big part of many state math standards. They can be used across multiple grades for a variety of concepts, and they help students understand and connect many related math situations. Understanding this model empowers students to create these representations as they encounter problems that they need to make sense of. The example in Figure 7.7 is a graphic organizer of a tape diagram that shows multiple representations of the equation $4 \times 5 = 20$. It could be included in the Reference Folder.

Figure 7.7 • Example of Upper Elementary Tape Diagram to Show Four Groups of Five

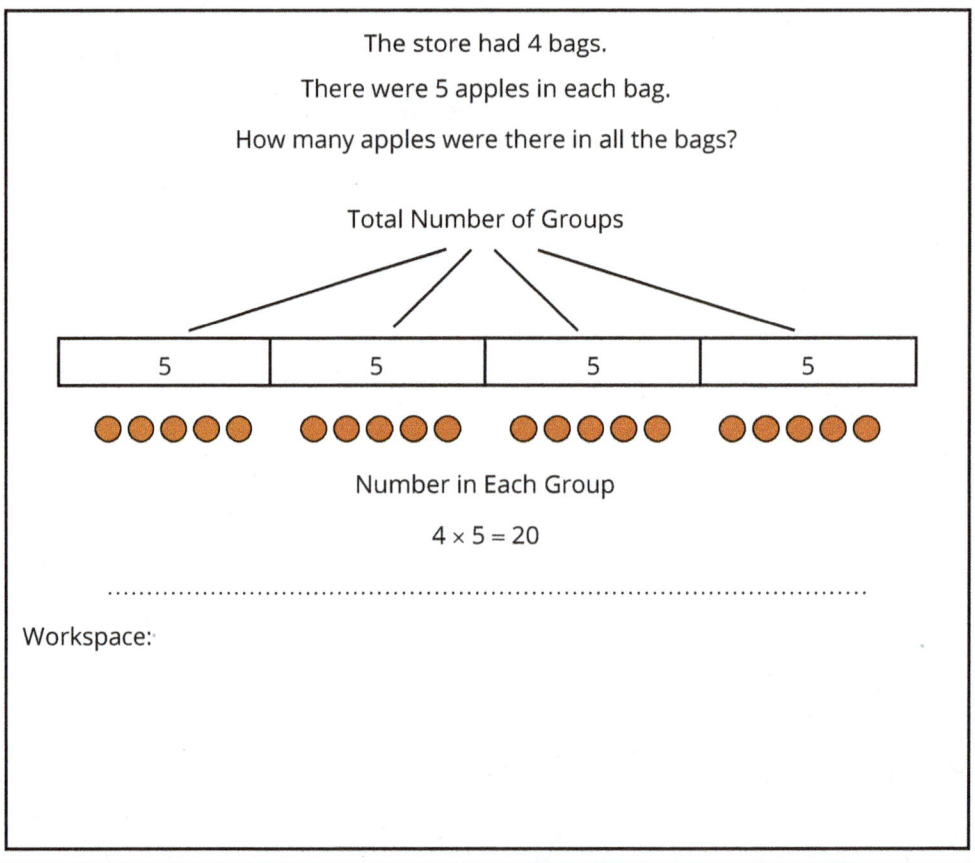

Source: Newton, 2024.

Other essential tools that support comprehension include "outlines, unit organizers, concept organizers and concept mastery routines" (CAST, 2024). Organizers can help students bridge prior knowledge with new knowledge and should be a part of every student's tool kit for the topic of study. For example, if students are studying fractions, they can have graphic organizers that lend themselves to comprehending big ideas with fractions. Effective organizers include a helpful structure and ideally a worked example, but they are open ended and completed by students. Figure 7.8 is an example with fractions. It asks this question: Grandma had $1\frac{1}{2}$ cups of sugar, and to make some hand pies, she needs $\frac{1}{4}$ cup of sugar for each hand pie. How many can she make? Note the drawings of the tape compare how many $\frac{1}{4}$ segments are in three $\frac{1}{2}$ segments.

Figure 7.8 • An Upper Elementary Example of a Tape Diagram to Divide $1\frac{1}{2}$ by $\frac{1}{4}$

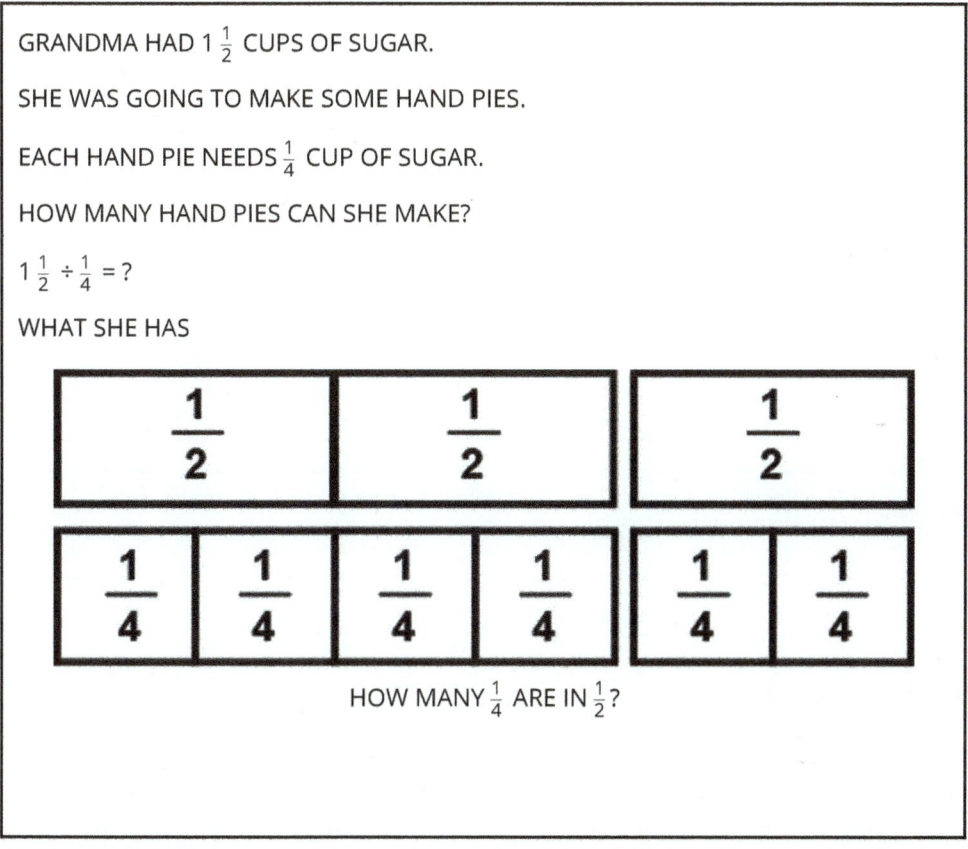

Source: Newton, 2024.

Interactive Graphic Organizers as Work Mats

Figure 7.9 is a nice example of a graphic organizer that creates structure but serves as an open-ended, interactive tool that students use to build understanding, make connections, and solve problems. The various models help students to organize their thinking about addition in more than one way. Work mats like these go in the Reference Folder and help students to organize their thinking in math workstations and Guided Math groups.

Figure 7.9 • An Interactive Graphic Organizer With a Number Line, a Ten Frame, and a Number Bond

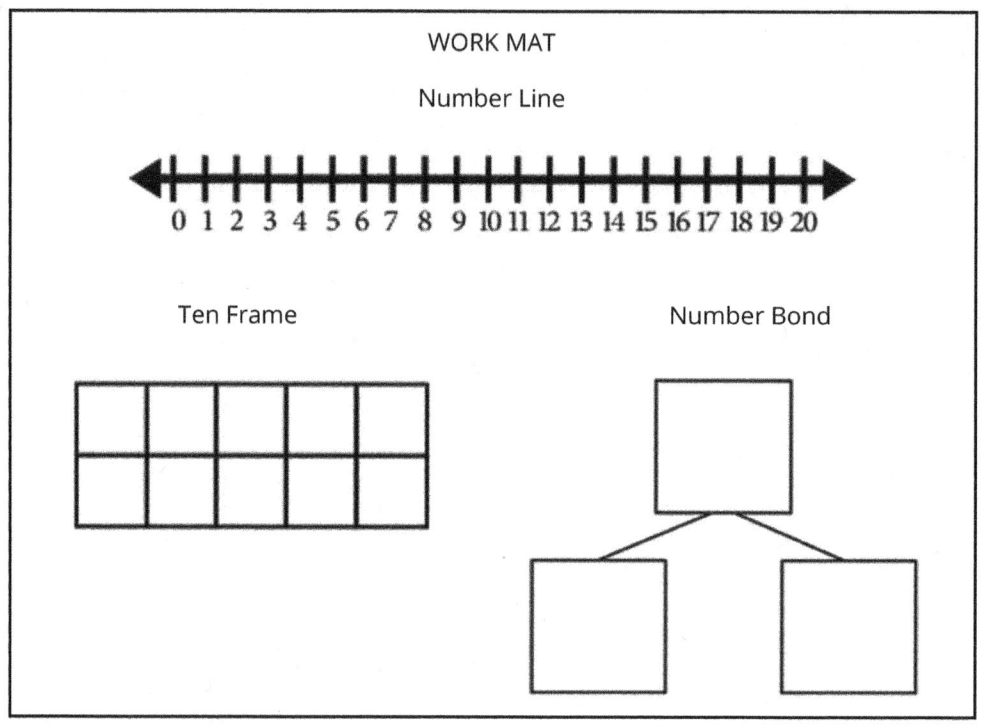

Download this Graphic Organizer at https://companion.corwin.com/courses/MathWorkshopPlus.

Triangle Vocabulary

Organizing language and text, and creating cues to support comprehension, is sometimes overlooked in math. Perhaps you are familiar with the Frayer model, which is helpful since it includes multiple representations. We offer an additional option for you as seen in Table 7.2. This example gives students vocabulary to describe different triangles and then asks them to write a definition or description and a drawing or sketch of each. As mentioned in Chapter 6, it is important to have various ways for students to organize and work with vocabulary throughout the unit of study. Students should always write the definitions in their own words. Students should draw examples as well. Remember, if barriers such as OT issues or stamina exist, offer technology to change the modality without changing the goal. Options such as typing on the computer, using speech to text, and adding clip art allow students to complete the task regardless of their ability to write or draw.

Table 7.2 • A Graphic Organizer for Triangle Vocabulary

Word	Drawing	Description
Scalene triangle		
Right triangle		
Equilateral triangle		
Obtuse triangle		
Acute triangle		
Isosceles triangle		

Unit Organizers

Unit organizers can be useful to keep students on track throughout the current unit of study. They frontload the trajectory of the unit and show students where they are going. This serves as compass for workstation activities because students understand the purpose and see what the ultimate goal is. Unit organizers can come in the form of checklists, organizers, sticky notes, and electronic reminders (CAST, 2024). Figure 7.10 is an example of a unit organizer for subtracting 2-digit numbers.

Figure 7.10 • Unit Organizer for Subtracting 2-Digit Numbers

Source: Newton, 2024.

Examples/Nonexamples

One of Alison's all-time favorite things to do is to use nonexamples, and she's always surprised that more people don't use them! Research has noted that the use of examples and nonexamples to help students understand a concept can deepen their comprehension (Bruner et al., 1956; Dean et al., 2012). There are many ways you can do this. You can lead students through a discussion where they look at examples and nonexamples and then have a conversation about what they notice and wonder. You can silently post examples on one side of the board and nonexamples on the other and have students share their conjectures about what's the same and what's different.

Another way you can explore the power of examples and nonexamples is to have students create their own with partners or small groups. Math Workshop Plus gives you multiple avenues to explore this. You might do it as part of your launch or during your mini lesson. Or perhaps you only want to do it with certain students in a Guided Math group, or maybe you want students to play around with it in workstations. Typically, it is much easier for students to name examples than to name nonexamples. Working on examples and nonexamples helps students to avoid overgeneralizing, undergeneralizing, or forming misconceptions (Malamed, n.d.). Providing organizers to support reasoning about examples and nonexamples is helpful and leaves students with documentation of their reasoning and analysis. Figures 7.11, 7.12, and 7.13 offer examples.

Figure 7.11 • An Example and Nonexample Organizer on Hexagons

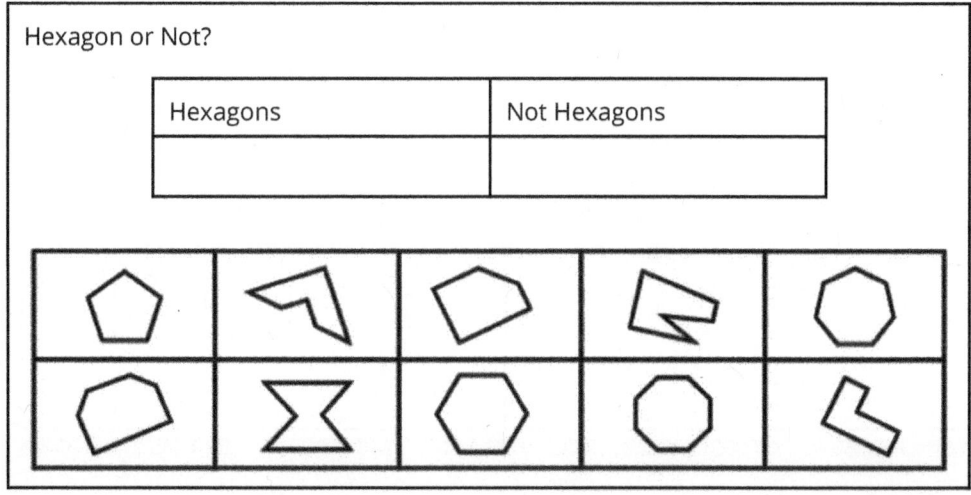

Figure 7.12 • A Frayer Model on Polygons

Definition A closed, plane figure with straight sides and angles	Characteristics Straight line segments No curved lines 2D figure More than two sides closed
Polygon	
Examples 	Nonexamples

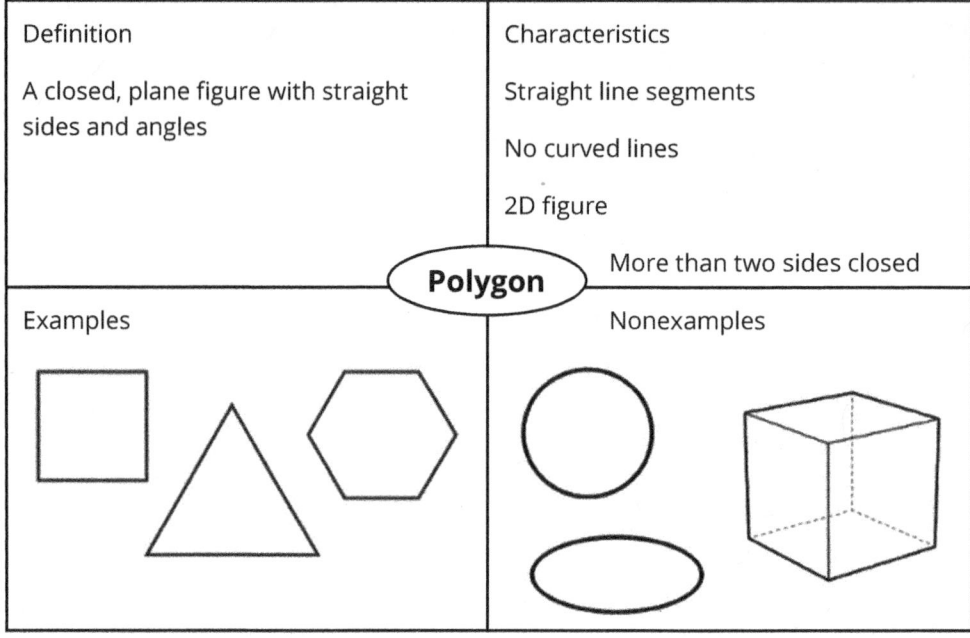

Figure 7.13 • A Frayer Model on Absolute Value

Definition The distance a number is from zero on number line **Absolute Value**	Characteristics The symbol for absolute value: $\vert\ \vert$ **It is always <u>Positive</u>**
Examples $\vert 5\vert = 5$ $\vert -3\vert = 3$	Nonexamples $-\frac{1}{4}$ **-14** **-8**

Cues and Prompts

Cues and prompts can help students successfully navigate mathematical problems. Prompts help students get started, remember what to do, reinforce big ideas, and explore different ways to approach problems. Cues and prompts serve as scaffolds to make math more accessible. Fisher and Frey (2014) note that there are four types of cues; chances are that you already use them:

1. Visual
2. Verbal
3. Gestural
4. Environmental

Visual cues are those you can see, such as highlights, stars, or bold text. Visual prompts may be included in slides, on anchor charts, or within graphic organizers. Verbal cues and prompts are spoken. They may be in the form of a statement, question, or call and response. For example, a child who is working on rounding might have a number line and you might cue a student to get started by asking, "What two numbers should we benchmark first?"

Gestural cues, such as those shared in Chapter 6, involve physically pointing, showing, or holding something for students to make meaning. For example, if you were teaching fractions, you might point to the denominator when talking about the whole or the total number or pieces and the numerator when talking about the number of pieces we are looking at.

Environmental cues involve using physical objects in the classroom or using manipulatives. For example, if you sent students on a hunt for rectangular prisms in the classroom and showed a tissue box to get them started, they would find lots of others. Within Math Workshop Plus, we intentionally create an environment that offers students cues. You could create an anchor chart that cues students to use their tools or clean up their workstation or be a good partner. We've seen teachers use photos to prompt students to return the tools neatly, showing them what the area should look like for the next students.

You may find yourself using some combination of these cues, which is excellent from a UDL perspective. For example, pairing your verbal cue with a visual amplifies the impact. Picture this. You are transitioning from the workstations or Guided Math groups to call everyone back for the debrief, and you may raise your hand (gestural cue) to get everyone quiet and then announce that it is time to come back together (verbal prompt). One of our favorite moves is to use a gestural cue when a student is about to interrupt a Guided Math group. Before they can make it that far, we simply point to the "When It's Okay to Interrupt" anchor chart (see Figure 7.14) to redirect them and send them on their way!

Figure 7.14 • Example of Guided Math Redirection Poster

Connections to SEL

Self-Management

Helping students to manage their learning through organization is an important goal of teaching. You provide the necessary scaffolds to process the information, manage their learning, and stay organized. This is an academic skill that intersects directly with self-management and involves explicit instruction. Students then learn these skills and use them, which requires self-reflection and decision-making.

> **Equity Check**
>
> Equitable Access
>
> As we look through an equity lens, we are always searching for ways to help students access information. In your Math Workshop Plus classroom, you will use graphic organizers and other instructional supports to highlight patterns and point out critical features, big ideas, and relationships. Since students access information in different ways, you will share information in different ways. As you use strategies to help students stay organized and focused, you help them build the capacity to learn to do these things on their own. When thinking about equitable access, you provide structures, systems, and explicit scaffolds that help students to succeed on their unique learning journey.

Cultivate Multiple Ways of Knowing and Making Meaning

For students to comprehend the different parts of the topic you're presenting in a unit of study, consider how to break apart that information into chunks that are easier to understand. There are many different ways to do this, always considering how to best chunk, prioritize, categorize, contextualize, and summarize information (CAST, 2024). Learners are in different places with these different skill sets, and it's important to consider where students are, the skill sets that they have (what *can* they do), and what they need to support their comprehension. Certain tools can foster independence during math workstation activities because they serve as "just in time" scaffolds (Dixon, 2018, para. 3). There are many different pieces that make this puzzle all come together, such as using various organizational methods, customizing and embedding interactive models, inserting scaffolds, and using feedback systems that support, engage, and motivate learners (CAST, 2024).

Provide Options for Organizational Methods and Approaches

There are several kinds of organizational charts and tables that help students process mathematical operations. For example, Figure 7.15 shows a cardinality chart used with primary students. A cardinality chart is a research-based tool that helps young children see the relationship between the number and representation of the number.

Figure 7.15 • A Cardinality Chart to Support Primary Grade Learners

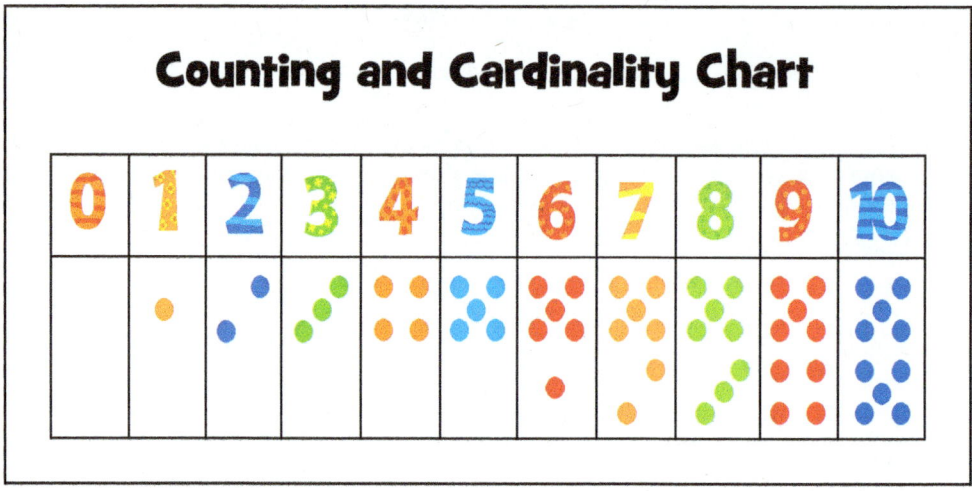

For upper elementary learners, multiplication wheels can be great tools because they provide an alternative to a multiplication table to find facts. Note that students should use these to *check* their answers, not *find* the answers when engaging in workstations (see Figure 7.16).

Figure 7.16 • Multiplication Wheels Chart to Support Upper Elementary Learners

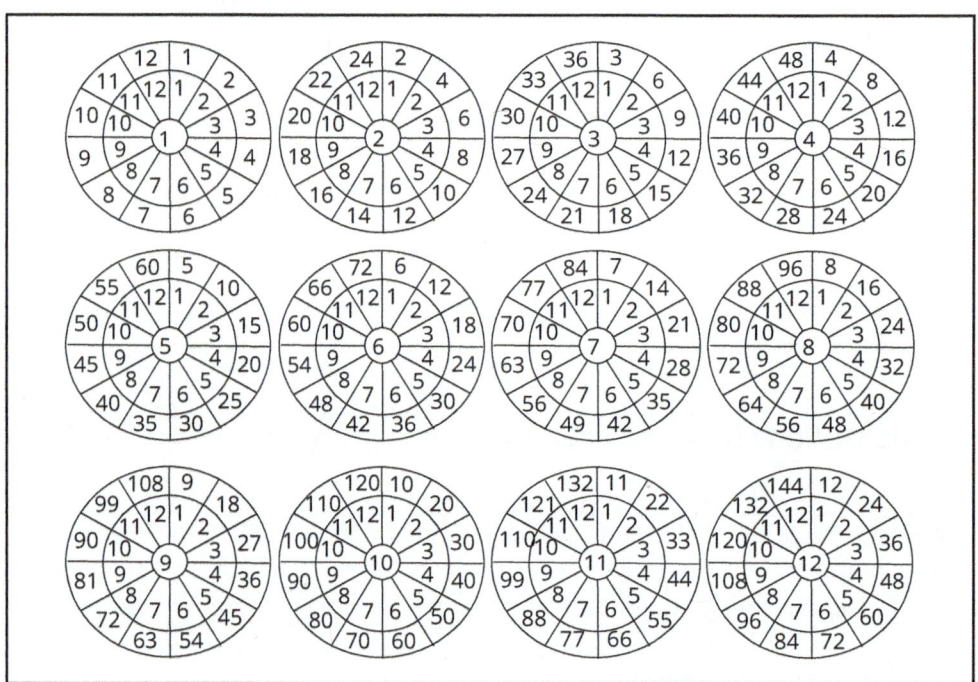

Upper elementary and middle school students can benefit from a prime number chart (see Figure 7.17). A prime number chart is a visual tool that helps students to see the patterns by highlighting prime numbers. The chart helps students to identify prime numbers quickly, observe how prime numbers are distributed, and explore patterns and number relationships such as twin primes.

Figure 7.17 • A Prime Number Chart to Support Upper Elementary and Middle School Learners

Provide Interactive Models That Guide Exploration and New Understandings

As we continue to say, virtual manipulatives are great tools that help students to explore math concepts. If you haven't tried them yet, give them a shot. As long as students have access to digital devices, they are a great option. Digital Rekenreks, such as the one in Figure 7.18 from Brainingcamp, are powerful tools to explore the operations within 20 and 100. In this example, students are working on the concept of "doubles plus 1" facts within 20. This visual tool allows students to see the math quite clearly.

Figure 7.18 • Digital Rekenrek

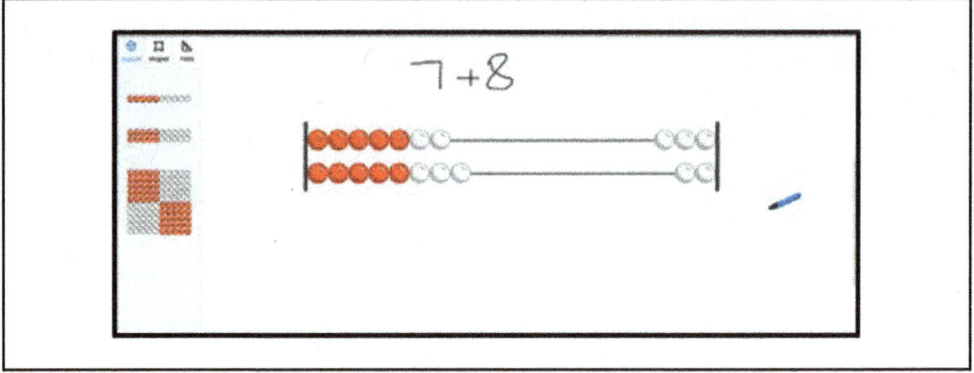

Source: Brainingcamp.

Operating with fractions is another concept that can be difficult for students to understand. Virtual number lines can help students illustrate their thinking, showing how they jumped up and back. The number lines can be set at different intervals and represent whatever values students are working with. Students can explore the different operations by moving up and down the number line and explaining their thinking. Figure 7.19 shows how students can explore the expression $3 \times \frac{1}{8}$ by jumping up by $\frac{1}{8}$ three times on a number line.

Figure 7.19 • A Digital Number Line for Fractions

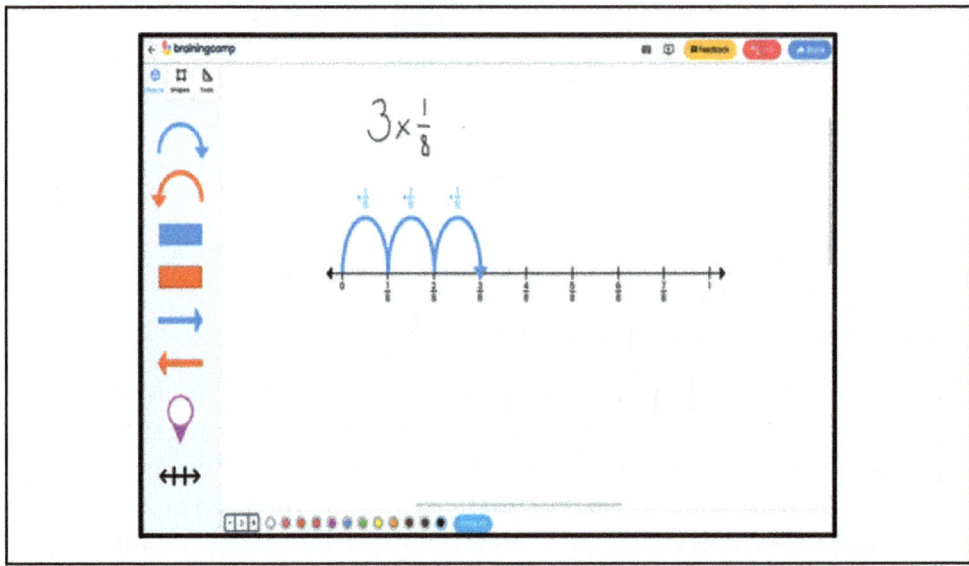

Source: Brainingcamp.

Using the percent manipulatives is a way of helping students build understanding of percentages through reasoning. As you can see in Figure 7.20, virtual manipulatives allow the students to easily convert between fractions, decimals, and percentages so they can make meaningful connections. If you favor grid paper for percentages, just remember that although these percent bars may not be your first choice for this, they may help some students unpack the topic in more than one way.

Figure 7.20 • Digital Percent Bars to Convert From Fractions to Percentages

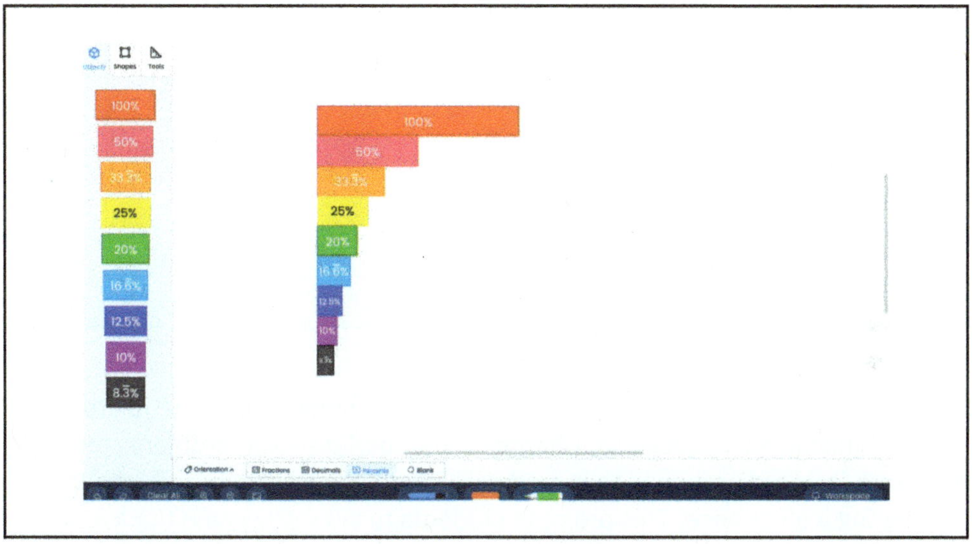

Source: Brainingcamp.

Introduce Graduated Scaffolds That Support Information Processing Strategies

Quiz time! Are students more likely to comprehend information if you tell them the information or if you ask questions to guide and activate their thinking? Questioning is the key to flipping the "on" switch in students' brains, and carefully crafted questions support students in making the connections needed to understand what's happening. For example, when you enter with simpler versions of problems first and then work up to more complex versions, the questions you ask along the way connect the dots between them. Breaking big pieces of information and content into smaller, more accessible—but connected—pieces so that students can digest them more slowly and carefully has been shown to be extremely effective (Liljedahl, 2020). Activities like these can be done during the whole-group portion of Math Workshop Plus and are critical during Guided Math groups where you use probing questions to draw out what students know and help them make connections.

There is a great deal of research showing that worked examples can be very helpful to students (Atkinson et al., 2000; McGinn et al., 2015). In Exhibit 7.2, students explore several scaffolded examples, with a little less scaffolding at each level. We call this a faded worked example. By the time they work on the fourth problem, students are expected to do it on their own.

Exhibit 7.2 • Faded Worked Example

> $2 \times 10 = (2 \times 5) + (2 \times 5)$
>
> $4 \times 10 = (4 \times \_) + (4 \times \_)$
>
> $5 \times 6 = (5 \times \_) + (\_ \times 3)$
>
> 8×7
>
> In a faded worked example, the scaffolding fades away slowly but surely. Eventually, students can solve the problem without any scaffolding.

Numberless word problems are a great example of chunking information to activate thinking. As discussed in Chapter 6, numberless word problems offer a structure for word problems that progressively releases information rather than sharing it all at once. This prompts students to slow down and think about the information they have before new information is added (Bushart, 2023). Numberless word problems are great to explore during your whole-group opportunities as well as your Guided Math groups because the discourse among students is powerful as students share their thinking and their strategies with each other. Exhibit 7.3 offers an example.

Exhibit 7.3 • Numberless Word Problems as a Form of Scaffolding

> 1. Kimi and her sister went to the beach. They found some shells.
> 2. Kimi and her sister went to the beach. They found some shells. *Kimi found 7 shells.*
> 3. Kimi and her sister went to the beach. They found some shells. Kimi found 7 shells. *Her sister found 5 shells.*
> 4. Kimi and her sister went to the beach. They found some shells. Kimi found 7 shells. Her sister found 5 shells. *How many shells did they find altogether?*

Connections to SEL

Self-Management

It is important to teach students how to think about what they know, what they are struggling with, and how to use that knowledge to become aware of what they need to successfully manage their learning. Self-awareness precedes self-management, so be aware that it may be a process to get certain students to do both. Students benefit from on-demand tools and templates that help them work through stumbling blocks they encounter. If you take on the responsibility of providing the templates and scaffolds, then students can take on the responsibility for choosing which ones work for them and using them as they want or need them.

Equity Check

Culturally Relevant Information Processing

As you plan for information processing and visualization, you are probably thinking about what individual students need to access and process the knowledge you plan to share. You may be thinking right now about who will benefit from what structure and how the right tools may unlock things for certain students. Hammond (2018) reminds you that to do this equitably you must take a culturally relevant approach to information processing. This means that you might process "new content using methods from oral traditions" or "connect[ing] new content to culturally relevant examples and metaphors from students' community and everyday lives" (p. 41).

An equitable approach also means thinking about the cognitive scaffolds needed for some students to access activities. This requires structures and systems that allow flexibility in presenting information, scaffolding information, and building content knowledge. You know your students. You know that with a faded worked example, a barrier may be removed for some, but not everyone needs that. Equity is not one-size-fits-all, and Math Workshop Plus should exemplify that.

Maximize Transfer and Generalization

We know that you want all learners to develop a deep understanding of the content so they can take that knowledge and use it in new and exciting ways. When you keep learner variability in mind, you get more comfortable with the fact that just because it will take some students a longer time than others to generalize and transfer knowledge, it doesn't mean they can't. The key is to create an environment that supports this. Students require different amounts and types of scaffolding for knowledge to go from short-term memory to long-term memory, and we have discussed so many strategies already. Yet we see the highest frustration levels among both teachers and students when confronted with problems and situations where students are expected to generalize and transfer their learning to new and unique situations. Infusing your Math Workshop with UDL supercharges your ability to provide opportunities for all students to encounter content in multiple ways so when it's time to apply it, they are ready!

Many of the strategies used to support comprehension can be used in different ways to support transfer. Word webs, graphic organizers, and concept maps can be used to introduce concepts as well as to review them. Interdisciplinary connections can be made throughout a unit that authentically engage students in applying what they know. This authenticity matters because it shifts students from "doing math to get an answer" to "using math to solve a problem." This is why we aren't really fans of many mnemonic strategies because many of them are based on gimmicks and tricks designed to get an answer rather than to make connections and promote deep understanding of mathematics. We do, however, love using visual imagery throughout a unit of study. Visuals can be powerful levers that can support comprehension from the initial learning experience and bridge it to transfer to the long-term application of the knowledge, which is the ultimate goal (Hess, 2023).

To maximize transfer and generalization, take advantage of your workstations. This is where you can incorporate explicit opportunities for review and practice supported with templates, graphic organizers, concept maps, note-taking scaffolds, and study guides. Remember that you may need to spend time during the First 20 Days to teach study skills. Research shows that students are often told to study but not taught how to do it effectively (Hattie et al., 1996; Thorpe, 2010).

Study Guides

What comes to your mind when you see the words *study guides*, and how should they be designed so that they are effective? There are many ways to make study guides. The key is that *students are actively creating the content* in the guides. When they do this, they are studying! You can help students craft a study guide during a mini lesson, workstation, or Guided Math group. The process of making the guide is a study opportunity that can be done with a partner, a group, or independently. Once they make the guides, you can lead a whole-class discussion about the big ideas to ensure everyone has the correct information.

After that, let your students know that you expect them to use their guide to study! You could make this a Must Do on your menu, and the artifact could be a note to themselves (or to you) about what was easy and what was tricky. You could even have students include a goal for the next day about how they will work on what was tricky, such as visiting a specific workstation or signing up for a Guided Math group to receive additional instruction.

In this first example, Table 7.3, you would walk students through the various examples for each big concept in the unit. You could give Example 1 and have the students give Example 2. Or you might review the big idea and then have students come up with both examples.

Table 7.3 • Study Guide for Fractions Example 1

Topic	Example 1	Example 2
Adding fractions with like denominators		
Subtracting fractions with like denominators		
Multiplying fractions		
Using different models		

Table 7.4 is another organizer for reviewing the big topics in a unit of study. In this example, the important topics are outlined, and students must fill in the information. You can decide if you want them to study in school with a study buddy or at home.

Table 7.4 • Study Guide for Fractions Example 2

Topic	Example
Adding fractions with like denominators	
Subtracting fractions with like denominators	
Adding mixed numbers	
Subtracting mixed numbers	

Figure 7.21 is one additional example of a study guide. Here, you include a visual model accompanied by language frames that students must complete and then use to study.

Figure 7.21 • **Study Guide for Fractions Example 3**

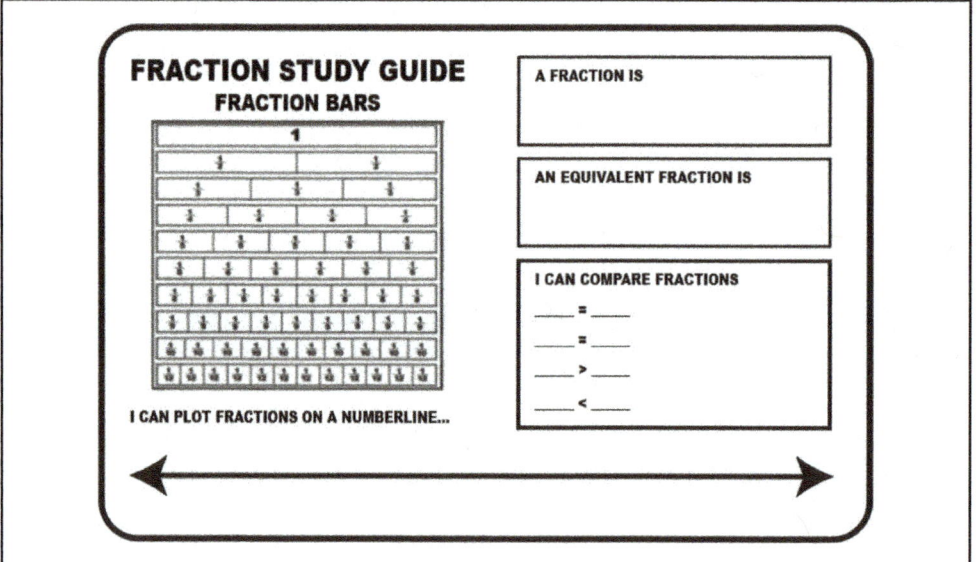

 **Connections to SEL**

Teaching Study Skills

Testing can cause students to become dysregulated. Teaching students what to expect and how to prepare can help avoid unfortunate moments of stress. Taking the time to scaffold the process throughout the year by teaching various approaches to studying builds self-awareness and increases student responsibility. When you normalize self-assessment along the way and offer informal and formal feedback prior to the test, students are less likely to be surprised by what the test expects of them. Doing this gives them the opportunity to make a plan to practice what they don't yet understand before they are assessed on them. Managing this themselves is not only empowering but builds that competency for future use. So while you may initially give students graphic organizers and study templates, eventually students will come to know using them is helpful, and therefore they seek them out on their own. Having and using study skills can help students to become "independent, self-regulated learners" (Walck-Shannon et al., 2021, para. 1; see also Smith, 2021).

Equity Check

Equity is about getting all students to make connections to their mental "file folders" and learning how to organize the new information in their brains so they can retrieve it and use it in new contexts. This is what experts do (Donovan & Bransford, 2005; Stern et al., 2021). When you commit to teaching with an equity lens, you instruct with the goal of transfer for all students. This drives you to connect everything to the big ideas, concepts, and frameworks rather than just teaching isolated, unrelated pieces of information (Stern et al., 2021). This changes lessons and standards from something we see out the window and are gone as we pass them by to a destination where we stop, gather observations and ideas, and connect them to the larger world that we live in.

Say Bye-Bye to Barriers When You . . .

- Tap into prior knowledge and connect to new learning
- Use advanced organizers
- Make cross-curricular connections
- Use examples and nonexamples
- Use cues and prompts
- Scaffold and chunk information
- Review and practice
- Cultivate multiple ways of knowing and making meaning
- Make connections beyond the text

Action Plan

Assess your current practice in terms of Designing Options for Building Knowledge. Download the Chapter 7 Self-Assessment at https://companion.corwin.com/courses/MathWorkshopPlus.

Try It! Supporting Comprehension

Identify three areas where you can tweak the ways in which you support comprehension. Try at least one of them and see how it works.

Connections

Identify a quote that stood out for you in this chapter. Post it on social media and share why it grabbed you. Don't forget to tag it with #mathworkshopPLUS!

Chapter 8

Design Options for Interaction

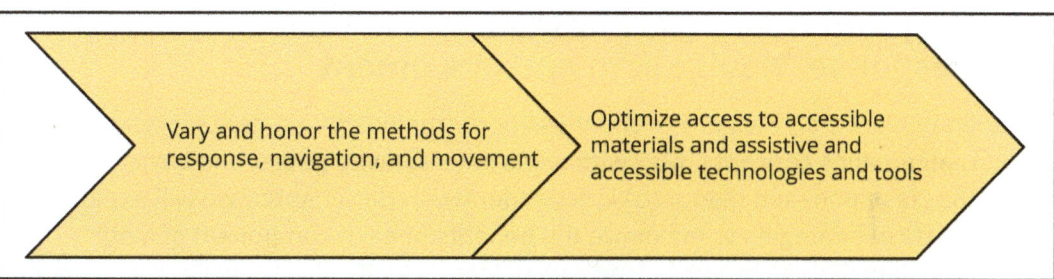

As we have discussed, students learn and process information in many ways, yet often our classrooms present information in ways that don't match the variability of learners (Meyer et al., 2014). This is true in terms of how we teach and how we assess. When you envision a traditional math lesson, you may picture students sitting at their desks, listening to the teacher talk, completing guided or independent practice, and demonstrating their understanding through tests or quizzes, some of which may even be timed. While effective for some students, these approaches can result in opportunity gaps for students who acquire and process information differently and who may need additional modalities to effectively show what they know.

In a Math Workshop Plus classroom, the focus is on accessibility and equity. As you've seen throughout this book, students move around the room, select and use tools, make choices, and have multiple and varied opportunities to demonstrate their understanding. This type of learning environment respects a variety of approaches, diminishes a focus on speed, and leverages technology to meet students in their Zone of Proximal Development (Vygotsky, 1962). In this chapter, we will examine each component of Math Workshop to uncover opportunities to increase and maximize *Design Options for Interaction*. The following questions will help you reflect on each element of your workshop, consider how to best vary the methods for response and navigation, and optimize access to tools and assistive technologies (CAST, 2024):

> 1. *Vary and honor the methods for response, navigation, and movement*: What options are available for students to show what they know and work at their own pace in a space that meets their needs?
> 2. *Optimize access to accessible materials and assistive and accessible technologies and tools*: What tools can students access, and do they know how to use them?

Vary and Honor the Methods for Response, Navigation, and Movement

Math Workshop Plus offers countless opportunities for students to explore mathematics concepts, play with ideas, interact with peers, and navigate the classroom and resources independently. In this chapter, we will explore practical strategies to maximize the benefits of each component of Math Workshop Plus to promote interaction. More specifically, we will zoom in on the launch, mini lesson, Guided Math groups and workstations, and the debrief to identify opportunities to do these things:

- Allow for variability in processing speed and the range of motor action required to interact with instructional materials, physical manipulatives, and technologies

- Provide options for physically responding or indicating selections

- Offer alternative options for physically interacting with materials by hand, voice, or keyboard

- Reimagine the physical space to include flexible seating and spaces which offer variable lighting, noise control, and so on

Adapted from CAST, 2024

Routines and Energizers: The Launch

We love launching Math Workshop with routines and energizers because, by design, they offer different ways to engage students and can provide opportunities for movement. These quick exercises get students involved and learning from each other as they interact. The open nature of the tasks naturally supports equity, accessibility, and the interaction needed to strengthen SEL (social-emotional learning) skills. If the thought of different routines and energizers feels overwhelming, we recommend sticking with Number Talks to establish a routine until you're ready to diversify.

Number Talks

Number Talks (Parrish, 2010; Sun et al., 2018) center student thinking and allow for various forms of expression. During a Number Talk, students solve problems mentally and initially communicate using only hand signals. Different signals, such as those shown in Figure 8.1, communicate different things, such as "I'm thinking" (closed fist), "I have a strategy" (thumb up), "I have a solution," and "I agree" (thumb and pinkie extended). Using these signals honors different processing speeds since students don't raise their hands as soon as they find a solution. Instead, the hand signals are shown below the chin in a subtle way that is not obvious to others. This avoids disrupting thinking while also silently communicating who's ready so you can provide adequate wait time. In this way, every student gets think time, and their needs are respected.

Figure 8.1 • Silent Hand Signals for Number Talks

Source: Dr. Nicki Newton

 Download this poster at https://companion.corwin.com/courses/MathWorkshopPlus.

Once students have a solution, they share their strategies with the group as you record their thinking on an anchor chart such as those in Figures 8.2 and 8.3. To explain their thinking and show how they arrived at the solution, students may use visuals, words, or other tools such as number lines. The anchor chart serves as visual artifact of the talk that can serve as a classroom tool for future reference and learning. At the end of the Number Talk, Alison loves asking students to review the anchor chart and share which strategy *they* found to be the most efficient, effective, and easy (the 3 Es), noting that this may be different for different people. This is a very public and visible affirmation that the routine honors and celebrates differences, and it validates different ways to arrive at solutions, something often missing in traditional math classrooms.

Figure 8.2 • Number Talk Anchor Chart Example 1

Figure 8.3 • Number Talk Anchor Chart Example 2

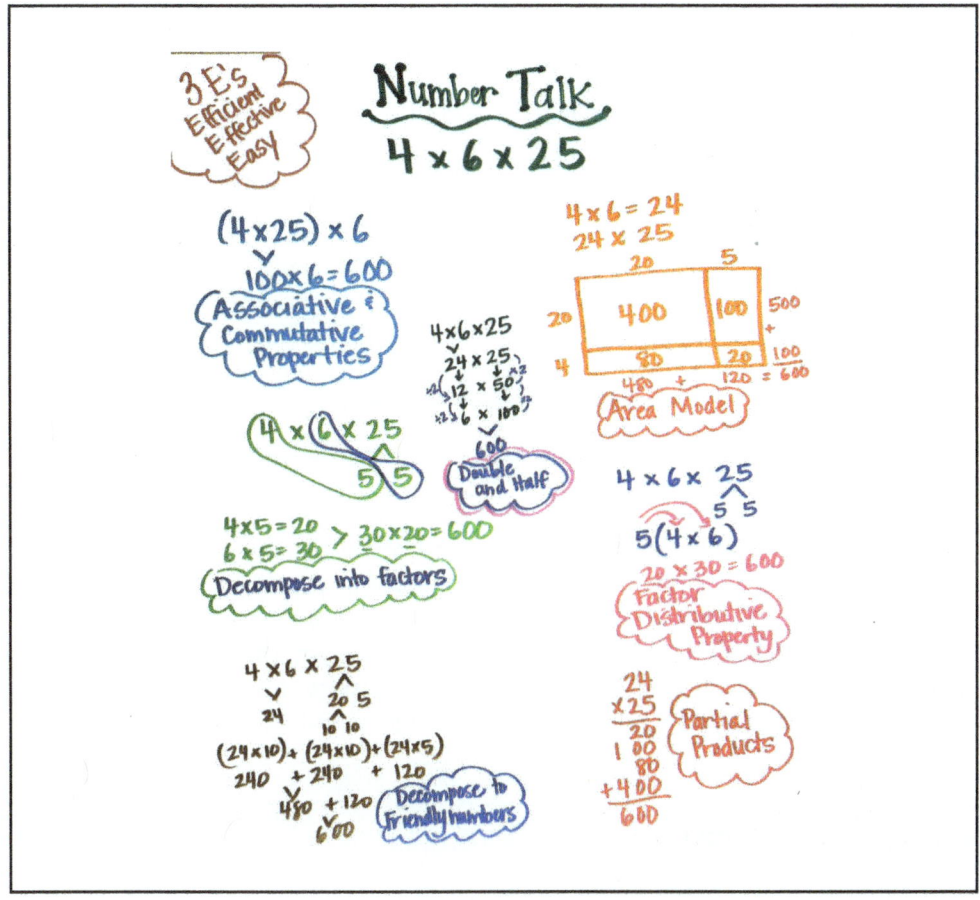

Table 8.1 shows some strategies to make Number Talks even more equitable and accessible by varying the methods for response and navigation.

Table 8.1 • Strategies for Making Number Talks Equitable

Strategy	Methods of Response and Navigation
Supporting communication	• Language frames • Individual whiteboards • Turn and talk vs. whole-group share • Visual vocabulary cards • Communication boards

(Continued)

(Continued)

Strategy	Methods of Response and Navigation
Integrating manipulatives	• Allow on-demand access to individual tool kits • Allow on-demand access to class manipulatives • Allow use of virtual manipulatives
Leveraging technology	• Use student response systems

Connections to SEL

Using Routines to Foster SEL Competencies

Beginning the math block with energizers and routines allows students to talk with each other, listen and learn from one another, and practice how to agree and respectfully disagree. It also requires students to wait for their turn to share, communicate succinctly, and not blurt out answers. This sets the tone for the workshop by offering authentic experiences for students to practice self-awareness, self-management, social awareness, and relationship skills.

Other Launch Routines

There are many math routines freely available online that include visuals, allow students to share thinking in various ways, and offer the "low floor–high ceiling" aspect. Two of our favorites are Esti-Mysteries by Steve Wyborney and Which One Doesn't Belong by Andrew Gael. Let's look at each of these in more detail.

Esti-Mysteries is a routine where students are shown an object and gradually given clues to work their way toward a solution. You can see examples for primary grades (Exhibit 8.1, Figure 8.4), upper elementary (Exhibit 8.2, Figure 8.5), and middle school (Exhibit 8.3, Figure 8.6).

Exhibit 8.1 • Primary Example of an Esti-Mystery

- Clue 1: The answer is greater than 20.
- Clue 2: The answer is less than 41.
- Clue 3: The answer is part of this pattern 22, 24, 26, ___.
- Clue 4: The answer does not include the digit 3.
- Clue 5: The answer is not 24, 26, or 28.

Figure 8.4 • Primary Esti-Mystery Showing a Jar Full of Tops, a Dice, and the Clues

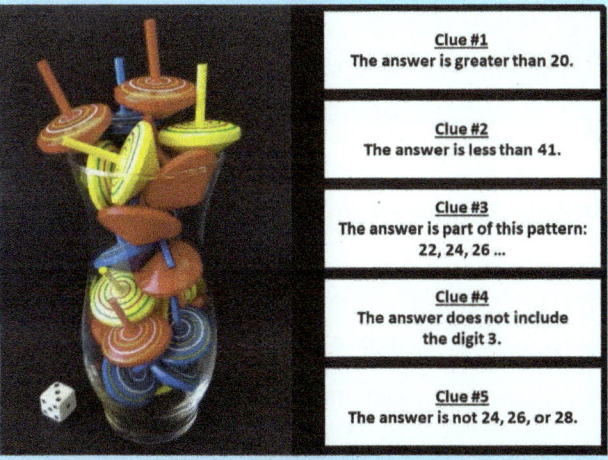

Exhibit 8.2 • Upper Elementary Example of an Esti-Mystery

- Clue 1: The answer is a multiple of the number of bananas you can see.
- Clue 2: The answer is not a multiple of 9.
- Clue 3: Count the bananas again. The answer does not include that digit.
- Clue 4: The answer does not include the digit 2 or the digit 3.
- Clue 5: Double 42 and eliminate that number.

(Continued)

(Continued)

Figure 8.5 • Upper Elementary Esti-Mystery Showing a Jar Full of Fruit-Shaped Candy and the Clues

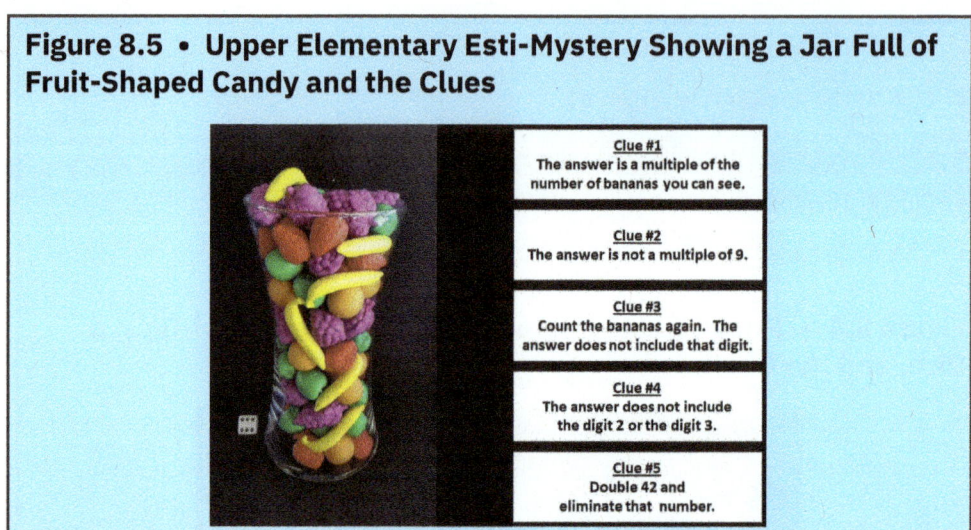

Exhibit 8.3 • Middle School Example of an Esti-Mystery

- Clue 1: The answer is a 2-digit number that is a multiple of 7, but it is not 84.
- Clue 2: The answer does not include the digit 1 or the digit 2.
- Clue 3: The answer is not a square number.
- Clue 4: One of the remaining numbers is half of another. Cross off both numbers.
- Clue 5: Eliminate the numbers in this pattern: 7, 21, 35 . . .

Figure 8.6 • Middle School Esti-Mystery Showing Two Plastic Cups of Beads

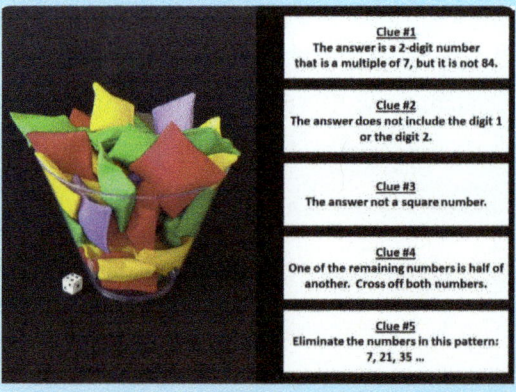

Source: Steve Wyborney

To vary the method of response and navigation for students, here are some ideas:

- Students record responses on a whiteboard.
- Students line up from least to greatest with each estimate.
- Have cards with the possible responses on them so students can just select a card.
- Use a student response system to allow students to use technology to participate.
- Offer language frames to support sharing.

Mini Lesson

When we think about the mini lesson component of Math Workshop, we envision different things. Depending on the day, topic, or the expectations in your context, you may have some combination of any of the following: explicit instruction (often the lesson in the math book/program); a read-aloud, song, or video; or an inquiry-based lesson that positions students to collaborate, make connections, and share conjectures (typically our personal choice). While all of these are acceptable and can be effective, it is important to consider barriers that may exist in each situation and remove them prior to engaging in the lesson. Let's explore each approach in Table 8.2 to identify potential barriers along with possible mitigation strategies.

Table 8.2 • Mini Lesson Types, Potential Barriers, and Solutions

Type of Mini Lesson	Potential Barriers	Strategies to Mitigate Barriers
Explicit instruction	<ul><li>Instruction is language heavy</li><li>Lesson driven by teacher modeling/telling</li><li>Lesson focused on procedures</li><li>Students expected to show proficiency in writing</li></ul>	<ul><li>Add visuals</li><li>Use questioning</li><li>Allow modeling with manipulatives such as base ten blocks, beaded or open number lines, place value charts, etc.</li><li>Allow students to draw pictures, use models, or use speech-to-text features</li></ul>

(Continued)

(Continued)

Type of Mini Lesson	Potential Barriers	Strategies to Mitigate Barriers
Read-aloud	• Auditory processing issues • Language • Speed of delivery too fast/slow • Students have trouble listening/attending • Comprehension	• Support concepts with visuals • Have students act out key parts or use gestures to illuminate concepts • Offer recorded version where speed may be adjusted • Allow students to access recorded version in chunks using headphones • Offer tools such as graphic organizers, word banks, pictures, and visual vocabulary • Use questioning to support comprehension
Song	• Memory • Tolerance for sound • Language	• Offer printed lyrics • Offer headphones to buffer • Have students track words on a screen or page as they sing
Video	• Comprehension • Language • Visual impairment	• Pause and use questioning to check for understanding • Turn on translation tools • Turn on captions • Utilize or create audio companion to visuals
Inquiry lesson	• Social skills • Language • Showing proficiency in writing	• Strategically group/pair students & model expected behaviors • Offer visual vocabulary cards, language frames, text reader, translation tools • Encourage use of manipulatives and/or drawings to model and solve as well as allow for oral explanations or speech to text

Connections to SEL

Using Mini Lessons to Foster SEL Competencies

As a reminder, it is a great idea to vary your approach to the mini lesson when you can so that the delivery does not become robotic for students. Mixing it up will keep students engaged, curious, and help maintain attention.

Different types of mini lessons offer different SEL challenges and opportunities. For example, if you have many students with lagging skills in social awareness and relationship skills, you may avoid the inquiry-based problem due to the collaborative dynamic involved. We encourage you to reframe this as an opportunity to *build those skills.* By taking the time to model and practice expected behaviors and adding supports such as language frames, this type of minilesson could help you reach more than just your math goals. If you decide to try this, it's helpful to offer an exit ticket that asks students to reflect on how they worked within the group. This can then be used to set goals (more on this in Chapter 10).

Figure 8.7 shows an exit ticket with a focus on SEL competencies. As you can see, a student is asked to circle what kind of partner/group member they were that day. They may choose from a variety of emojis to convey what their energy was as a partner that day. They then must circle what went well:

- I listened.
- I was kind.
- I took turns.
- I shared my thinking.
- I made good choices.
- I tried my best.
- I helped my group.
- I used tools appropriately.
- I didn't get frustrated.
- I was focused the whole time.
- I let others talk.

Figure 8.7 • Sample Primary Exit Tickets

(Continued)

(Continued)

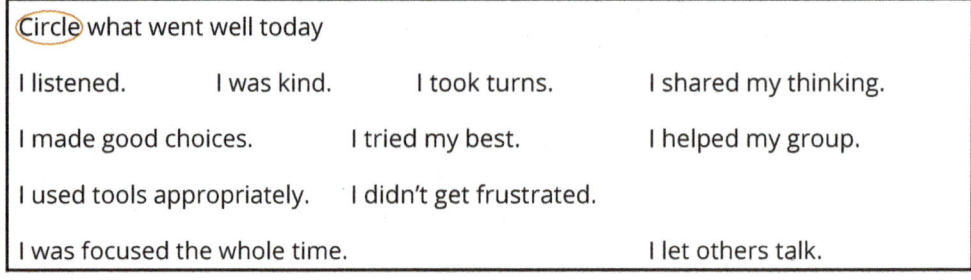

Source: Mello, 2024. Icons by Istock.com/lefym Turkin.

Figure 8.8 is designed more for upper elementary and middle school learners. It works on a scale of 1 to 5 as follows:

Figure 8.8 • Upper Elementary or Middle School Exit Ticket

5	4	3	2	1
I was fully focused the **whole** time. I gave my **best efforts** and worked with myself and my group.	I was fully focused **most** of the time. I gave **great effort** and worked pretty well with my group.	I got a bit off-task but regained focus by myself. I worked okay with my group, but I could have _____ better.	The teacher had to remind me to get on track. I struggled a bit in my group with (check all that applies) ☐ Listening ☐ Working together ☐ Taking turns ☐ Participating ☐ Being respectful	I didn't do my best work. I was off task too often. I struggled a lot in my group with (check what applies) ☐ Listening ☐ Working together ☐ Taking turns ☐ Participating ☐ Being respectful

Workstations

One of the easiest places to observe the intersections of UDL, SEL, and equity during Math Workshop is in the workstations. While you are meeting with Guided Math groups, the other students are engaging in workstations either independently, with a partner, or with a group, usually a combination of all three. Workstations offer authentic and powerful opportunities for students to strengthen their SEL competencies as they are charged with growing the skills in Table 8.3.

Table 8.3 • Skills Built in Workstations

Action	SEL Skill	Examples
Making choices	Responsible decision-making	• Good-fit partners • Just-right activities • Appropriate actions
Working with others	Relationship skills	• Being a good partner • Sharing space and resources • Being a good winner/loser
Staying on task	Self-management	• Setting goals • Using tools • Staying motivated and focused • Making the best use of time
Staying regulated	Self-awareness	• Managing emotions • Having a growth mindset • Persevering

Equity Check

Using Choice as a Lever for Equity

Workstations foster equity since all students are empowered to make choices that best meet *their* needs. As noted by Dabrowski and Marshall (2018),

Choice in process acknowledges and embraces the reality that all people, including children and adolescents, learn in different ways. Some people need social interaction to process and refine ideas, while others need internal reflection to find meaning. Some may take a linear approach to a task, while others may work in more recursive ways. Honoring these preferences means offering assignments in which students are given the freedom to design their course of action, sequence their steps as they go along, work alone or with peers, or manage timelines and deliverables. (p. 2)

If you prefer a station rotation over a menu or playlist, consider how you can offer choice within each station. This can empower students to self-differentiate and build agency. Choice can be in the form of math tools, partners, or different activities within the same station. These choices allow students to focus on their personal goals, help sustain attention for the duration of the rotation, and ensure they get what they need during those minutes.

Regardless of whether you use rotations, menus, or playlists, workstations must be independent. To make this a reality, take time to anticipate where students may encounter barriers, do what you can to remove those barriers, and model what options students have if they get stuck. Since this chapter is focused on varying the method for response and navigation, think about how this applies to workstations and how this impacts equity, access, and opportunity for success.

Equity Check

Ensuring More Than One Way

Consider the means of expression available in your current workstations. Is there more than one way for students to capture their learning at each station? For example, if you use a paper and pencil accountability sheet, are there students for whom that is not the ideal modality? Take a moment to think about your workstations. Do you have a mix of modalities? Are language supports included if oral or written communication is expected? Do alternative options exist for those who may need them?

Artifacts and Accountability "Sheets"

Student accountability is a vital component of successful workstations. Since students are empowered to make decisions about how they spend their math minutes during choice time, you will need systems of accountability to ensure that minutes are used purposefully. Teachers have many different strategies to collect artifacts from workstations, but most are not universally designed and tend to play to a single modality. Consider using some combination of the options in Table 8.4 to ensure that all students can effectively capture evidence of their learning while moving through workstations.

Table 8.4 • Examples of Options for Accountability and Artifacts

Primary Examples	Upper Elementary Examples	Middle School Examples
• Offer visuals to circle for nonreaders • Minimize writing by using cloze-type sentences • Allow students to use a notebook versus separate accountability sheets • Use Seesaw or other tools to collect evidence of work digitally • Limit the artifact expectation to one station, or recording one thing from each station	• Utilize digital tools such as Seesaw or OneNote Notebook to record work • Create a space in Google Classroom • Allow students to share their artifacts and explain their thinking through digital recording applications such as Padlet • Allow students to use the camera feature to take photos of their work and save them to a shared drive or folder	• Utilize digital tools such as OneNote Notebook or Schoology to record work • Create a space in Google Classroom • Allow students to share their artifacts and explain their thinking through digital recording platforms such as Padlet • Collect evidence of learning through applications such as EdPuzzle at workstations • Allow students to use the camera feature to take photos of their work and save them to a shared drive or folder

Purposeful Practice Activities

Using workstations to revisit the standards where mastery is essential is one way to make them purposeful. Workstations are meant to be intentional, differentiated, and targeted to a specific skill. When students select their workstations, they should do so with intentionality so that they get the practice they need for the specific skills they need. If this feels overwhelming, remember that technology can take much of the planning off your plate, and self-checking stations mean that there is no correcting for you! Refer to Exhibit 8.4 for some examples to get you started.

Exhibit 8.4 • Strategies to Ensure Purposeful Practice for All Students

- Leverage adaptive learning platforms such as Freckle, Dreambox, Boddle, Nearpod, etc.
- Offer varied formats and modalities for practice:
  - Games: individual, partner, group; digital, hands-on
  - Problem-solving: paper (add visuals and language supports); computer-based (utilize accessibility features such as screen readers)
  - Task cards
- Ensure immediate feedback
  - QR codes
  - Answer keys
  - Partner check
  - Self-checking activities
- Math Power Towers*
- Start/Finish Sticks*
- Pick/Clip/Flip game*

See supplemental materials for samples.

Student Communication

For workstations to run smoothly, communication between students is essential. To support this outcome, think about what may need to be added to your learning environment. For example, you could include language prompts within games or partner and group activities. Exhibit 8.5 offers some examples.

Exhibit 8.5 • Supporting Student Communication During Workstation Time

- Post language frames around the room.
- Create anchor charts with expected language during workstations and show students how to use it as a tool.
- "Good game!"
- "I can prove my answer by _____."
- "Can you explain your thinking to me?"
- "Can you help me?"
- Role-play scenarios and make an anchor chart for reference.

Another way to support interaction is by allowing students to decide where they want to engage in workstations. Some students prefer to stand, while others may opt to sit on the floor. Rarely are students in assigned seats during workstations, but you'll need to decide what you are comfortable with, what the boundaries are, and explicitly share those with students. For example, are you comfortable with students sitting on top of their desks? Would you allow students to work in the hallway? Can they use things in your room, such as bean bag chairs, scoop seats, or lap desks? These may seem like small things, but they can foster more equitable conditions if they make a difference in students' ability to focus, persevere, and remain on task.

Connections to SEL

When Lagging Skills Disrupt Workstations

When students have autonomy to move through workstations, they can operate in ways that naturally match their processing speeds (vs. a timer, for example), seek tools that work for them, and get as much or as little movement as they need. While these things help ensure equity and accessibility, this is a spot where lagging SEL skills show themselves loud and clear. Because students are tasked with self-management, responsible decision-making, and navigating relationships with partners and groups, you may experience some bumps. Taking the First 20 Days to establish the expected behaviors, model examples and nonexamples, and highlight the tools and supports available to students should help to minimize frustrations, build self-awareness, and help students to know where they are strong and where they need to grow. Resetting when needed throughout the year and stopping to take advantage of teachable moments to model appropriate actions and responses are powerful ways to build capacity in these areas.

Guided Math Groups

Small, flexible groups are a hallmark of Math Workshop Plus. As noted by Doubet (2022), "Groups are *flexible* because they align with specific, changing goals, and because decisions about group size, membership, and longevity are guided by recent classroom assessment results or other student or class characteristics relevant to a specific purpose" (p. 11). The design of the groups inherently removes some of the barriers that exist in the whole group, but even in this setting, you must consider what options for physical response and navigation you can include and how you can infuse tools and assistive technology when needed.

>  **Connections to SEL**
>
> Forming Guided Math Groups with Nonacademic Data
>
> When you call two to five students into a small group, they may all need to develop the same concept, but they may not all process and acquire information the same way. For this reason, it is helpful to offer choice to students, even within this small setting. For example, you can ask students which manipulatives they would like to use and if they prefer to use the physical objects or the virtual representations. Students can also work together or alone to act out problems, using themselves as the tool. They may share their ideas out loud, through a picture, in writing, or using their manipulatives. To support communication between students, try the table tent language frames from Chapter 6 or provide translation tools. We recently found translation earbuds for around $30 that work great!

If your school collects SEL data as part of a universal screening process or throughout the year, it is helpful to occasionally leverage that data to plan small groups and goals. For example, if certain students have relative strengths in self-management but relative deficits in self-awareness and/or relationship skills, you can use that information to form a Guided Math group. In this instance, you might choose to have students play a game during small group so they can practice winning, losing, and taking turns with you as a guide on the side. Prior to starting the game, you may want to review both the math concept and the expected partner behaviors. Through this Guided Math experience, you can simultaneously build math skills and SEL competencies.

Keep in mind that Guided Math groups need not happen at a "kidney table." Small-group instruction can happen anywhere, using any modality. In primary grades, consider having students jump on a number path or line up in consecutive order. Students with mobility or communication limitations can use a pointer or say where they would go. In middle school, Guided Math groups can meet at the interactive board to build understanding of integer operations using virtual tools such as red and yellow counters, work at their desks with the teacher using integer beaded number lines (red = negative/yellow = positive), or move to the back of the room or use the hallway to act out scenarios by walking a number line. Having each of these options increases the likelihood that all students will have the opportunity to participate and show what they know.

Equity Check

Independence Opportunity Gaps

While it may feel uncomfortable not meeting with every student every day, especially learners who tend to need extra support, consider the following questions: How often do striving learners get to work *without an adult*? Do striving learners have the same opportunities as their peers to play games, use technology, or work collaboratively on problems with peers? How often do students who quickly grasp concepts get to work with you? Remember that to fulfill the promise of equity and inclusivity in your Math Workshop Plus, you set the conditions for success for all students and ensure that opportunity gaps are eradicated. Giving yourself permission to not see every student every day will not only take some pressure off but will ensure that *all* students can use their strengths, learn to persevere, and truly own their learning without being dependent on you.

The Debrief

The debrief, or share, that closes out Math Workshop is incredibly valuable to the learning community but often is the piece that gets dropped when time is tight. We strongly recommend that you honor this time to provide closure, recognize and celebrate the learning of the day, and reinforce the expectation of accountability and personal responsibility.

There are countless ways to execute the debrief, and this is an important place to vary the methods of response to ensure that all students can share. Table 8.5 outlines three examples of how to conduct this portion of your workshop along with examples (Figures 8.9, 8.10, and 8.11). Each can offer different modalities and capitalize on different student strengths. If this is not currently part of your practice, know that the following strategies are designed to take no more than 2 to 5 minutes and do not require much preparation.

Table 8.5 • Sample Debrief Structures and Strategies to Vary the Methods of Response

	This, That, or the Other	**Jumps and Bumps**	**Journal Entry**
What to do	Pose different outcomes to the class and ask students which ones apply to them. Who learned *this*? Who learned *that*? Who learned something else . . . what was it? This may also related to degree of effort or understanding.	Ask each student to reflect on workshop and share a "jump" (a success) or a "bump" (a challenge).	Have students record their learning of the day in a "journal."
Strategies to vary the method of response	• Use a digital student response system. • Number the options and have students show which number they select by holding up their fingers. • Have students complete the activity in groups of three or four and discuss. • Collect the responses without names and arrange them like a bar graph or Concensogram. Have students discuss the trends. • Have students write their names on sticky notes and place them on a chart next to the outcomes.	• Have students write their jump or bump on an individual dry erase board. • Provide a pre-made template for students to fill in. • Create a digital form and have students submit their jumps and bumps using their device. • Offer students the option to create a 30-second video of their share using an application such as Flipgrid or Seesaw.	• Offer speech to text to complete the entry on the computer. • Allow students to create an audio entry by recording themselves with a microphone. • Provide pre-made rating scales or emojis to paste into the journal. • Have students create an ongoing video journal with 20- to 30-second entries. • Provide a pre-made list of possible responses and allow students the choice to circle one or create their own.

	This, That, or the Other	Jumps and Bumps	Journal Entry
Examples	**Figure 8.9** • This, That, or The Other Poster	**Figure 8.10** • Jumps and Bumps Poster	**Figure 8.11** • Journal Entry Poster

Connections to SEL

Debrief as Formative Assessment

The debrief should be a time for all students to celebrate successes and share their challenges. Keep an eye out for those who rarely share or who consistently highlight "bumps" rather than "jumps." These students may be struggling with confidence, motivation, or with social interaction and may need additional supports or different goals.

Equity Check

Ensuring That the Debrief Is Accessible

The examples are intended to highlight a variety of modalities for the debrief. If you tend to run the debrief the same way each day, consider who may benefit from a different opportunity. Some students may be missing out on fully sharing their learning if the structure has inherent barriers. If students don't have the means to participate, we don't have a true community of learners.

Optimize Access to Accessible Materials and Assistive and Accessible Technologies and Tools

Assistive technology, as defined in the Individuals with Disabilities Education Act (IDEA), is any item, piece of equipment, or product system that is used to increase, maintain, or improve the functional capabilities of the child (Lipkin & Okamoto, 2015). When you hear this, you might think of communication devices, FM systems, screen readers, or other items often prescribed through individual education plans (IEPs) for students who receive special education services. While all of those apply, through the lens of Math Workshop Plus we are thinking specifically about removing barriers that get in the way of math understanding.

What Are Your Barriers to Optimizing Access to Tools?

Think about your classroom. What tools do students use? We previously discussed that in many elementary classrooms there are least some math manipulatives such as counters, pattern blocks, base-ten blocks, and fraction bars. Depending on your grade level, you may have ten frames, counting bears, snap cubes, and attribute blocks, or maybe you have geoboards, algebra tiles, protractors, compasses, calculators, and rulers. Maybe you *wish* you had these things, especially since you are learning new ways to integrate these tools. Even if you only have paper for folding, 2-color counters, or dusty pattern blocks, where are they? Are they in a closet or cabinet or some other mysterious abyss? Are they in bins? Are the lids on the bins?

If you want to optimize access to tools, you need a plan and a system. Some teachers opt to create a designated area in their classroom where students can access the tools at any time. If you're less inclined to do this, consider why you feel this way. Though many of us struggle to release control, if you are this far into a book on Math Workshop Plus, that is likely not the reason. Over the years, we have frequently heard that teachers store things away or don't often use manipulatives because they worry about these things:

- Students will "play" with them.
- Students will use them inappropriately, possibly as weapons.
- If left out, the tools will disappear.
- The tools are distracting.
- The movement to get the tools is distracting.
- They take too much time to pass them out and put them back.

While these things are possible, we've found that using the First 20 Days to set clear expectations, model appropriate use, create anchor charts, and establish accountability are the key to making the "salad bar" model work. Some teachers prefer to distribute manipulatives and other tools as needed for specific lessons. While this is better than not having tools at all, it doesn't optimize access, foster agency, or promote independence. UDL is about accessibility, choice, and removing barriers. We have watched students who were frustrated and exasperated with missing addend problems figure them out easily with a Rekenrek or ten frame. The same is true with tools that support fraction and integer operations. When students can only access these supports when *you decide* to distribute them, they may believe they aren't good at math, when they learn best through multimodal experiences. Math Workshop Plus commits to ensuring students can provide this for themselves because the tools are always accessible. Figures 8.12, 8.13, and 8.14 will give you some ideas.

Figure 8.12 • Salad Bar Style Tool Depot

Figure 8.13 • Individual Math Tool Kits

Figure 8.14 • Access to Virtual Manipulatives

If you have class sets of manipulatives, it's ideal to give students personal tool kits. Keep in mind, though, that even with individual tool kits, students could be missing what they need. For example, if a student has issues with visual discrimination, executive function, or fine motor manipulation, virtual manipulatives could be a better fit. The bottom line is that one size won't fit all, so having a variety of options is key.

> ## Connections to SEL
>
> Normalizing Tools
>
> Optimizing access to tools and technology is one way to normalize them. Sometimes *certain* students are offered tools, or tools are only given during Guided Math groups. While well-intended, this may create an unintentional stigma and result in feelings of shame, perceived incompetence, and negative math dispositions. Normalizing tools signals that they are part of learning for all students. Moreover, we honor the element of choice that is a major component of both UDL and Math Workshop.

As an aside, tools are only useful to those who know how to use them, and that includes you! If you are unsure about how to use certain manipulatives, technology, or other tools, help is available. There are videos online for almost anything you can imagine. From learning what a Rekenrek is to learning about algebra tiles to finding how to turn on subtitles or translations when showing a video, if you want to learn, a quick search of the internet should do the trick. Once you have a handle on these things, you can teach your students. So many people assume that students know how to use the manipulatives or the accessibility features on their computers, and they don't. For this reason, we recommend explicitly teaching students how to use these tools and creating anchor charts that help students see when and how they could support their learning.

Optimizing Access to Technology

As noted in previous chapters, you likely won't be able to meet with every student every day. On days where you do not pull students into a Guided Math group, consider how technology may be used to offer an additional opportunity for students who may need extra support following the mini lesson. Whether it is part of a Must Do or a choice on a menu or playlist, technology options can allow students to revisit topics multiple times, on demand, and at their level—something that we cannot duplicate even in a small-group setting.

For example, some students may learn best when content is presented visually. For these students, you could offer choices that include videos or animations that utilize visuals. Some students may do better with auditory presentations. For these students, a listening station would be ideal so that they can listen, take notes, and replay any ideas that need to be repeated. If you are using a playlist or a menu, students can revisit these links as often as needed. If you have access to a platform such as Seesaw, Google Classroom, or Microsoft 365, think about how you can personalize the student experience by sharing resources that best match the preferred methods of response and navigation of your students. This will not only make students more successful but more independent. To continue to maximize opportunities for interaction and discourse, you may choose to allow students to engage with these resources in partners or small groups as part of their workstation experience.

> **Say Bye-Bye to Barriers When You . . .**
>
> - Embrace many ways for students to show what they know
> - Use technology to create access
> - Make manipulatives accessible and on demand
> - Remember that virtual manipulatives can help with management
> - Think outside of the box to upgrade routines
> - Offer a variety of ways for students to respond

Action Plan

Assess your current practice in Design Options for Interaction. Download the Chapter 8 Self-Assessment at https://companion.corwin.com/courses/MathWorkshopPlus.

Try It! Varying Your Responses

Identify three areas where you can vary the means of response. Try one of them and see how it works.

Connections

Create a system to make your math tools accessible. Share your photo on social media with #MathWorkshopPlus. Let's build the MWP community!

Chapter 9

Design Options for Expression and Communication

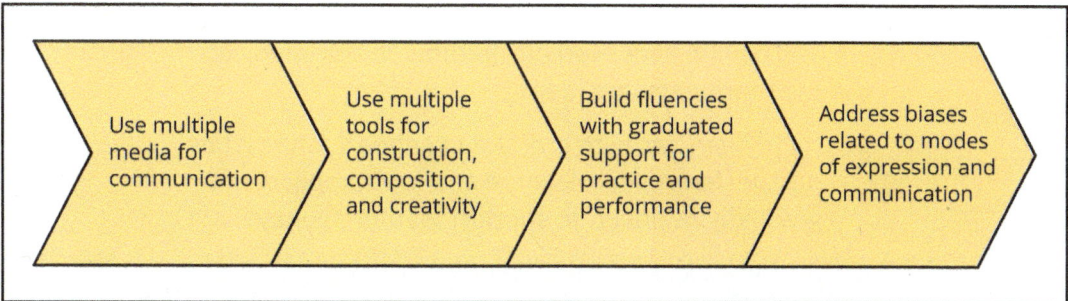

Throughout this book, we've explored countless ways to design an inclusive math experience that considers and respects the whole child. This chapter focuses on one of our favorite ideas: supporting students as producers and creators of mathematical ideas and imagining new ways for students to demonstrate their learning. If it feels like we already covered this, you've been paying attention! As you approach the end of this book, we hope you're making connections, seeing overlap, and feeling like this is an elevated version of what you already do, perhaps with more intentionality and a heightened focus on access and equity.

Since variations of some of the strategies and examples in this chapter have been shared in other chapters, albeit in other contexts, we ask you to shift your purpose here. As you read, circle things you *already* do, put a star on things you want to *start* doing, and jot down or cross out things you want to *stop* doing. Determine your role in supporting communication and expression, what immediate actions you'll take, and note which students may respond differently when you offer new ways to demonstrate proficiency. Think about the things you'll add, modify, or eliminate to elevate your practice to Math Workshop Plus.

This chapter may feel a bit like Chapter 8. While they are both about designing multiple means of action and expression, the purpose is different. Chapter 8 was more about the materials and physical environment; this chapter is more about diversifying modes of communication so all students can effectively communicate what they know. Use the following questions to reflect independently or with your team. As you explore the dimension of UDL that frames this chapter, which elements of *Design Options for Expression & Communication* already exist? Which need to be added?

> 1. *Use multiple media for communication*: What options are offered to students to express understanding in flexible ways?
>
> 2. *Use multiple tools for construction, composition, and creativity*: Which tools complement the learning goal and can be accessed by students to communicate their thoughts and ideas?
>
> 3. *Build fluencies with graduated support for practice and performance*: How are scaffolds applied and gradually released to support learner agency?
>
> 4. *Address biases related to modes of expression and communication*: In what ways can less common methods of communication be honored?

Think about the things within your control, not systems that present barriers (although we recommend bringing those to the attention of administration). For example, if you're thinking you can't use technology or hire someone to provide sign language because your school can't fund them, think beyond that barrier and shift to what you *can* use. We have seen some incredible learning come from simply folding and ripping paper!

Reflect on things you do "because you've always done them." Think about current practices that work, those that need upgrading, and those that probably should've been eliminated long ago. Think about tools and practices you haven't tried yet, and how they could help students communicate their learning. Enter this chapter with curiosity and look for ideas to make your own. And don't forget your pen to note what you want to start (★), stop (**x**), and continue (◯) within your practice.

> **Equity Check**
>
> Stretching Yourself
>
> Trying new things such as technology or manipulatives can be scary and intimidating. If you're reluctant, try baby steps. Not everything new is hard. Keep your focus on the goal of creating a more equitable learning environment and creating access to more students. Commit to trying something tomorrow. Tell students that *you're* learning, share why, and model that it's okay to make mistakes. We're confident your students will cheer you on!

Use Multiple Media for Communication

In Chapters 6 and 8, we explored using different media to illustrate concepts and ideas. We hope those examples showed how reducing the role of language allows access to big ideas through visuals and how different students benefit from different tools. Think now about the free or common media you likely have access to and how students use them, or could use them, to demonstrate understanding. Later in this chapter, we'll revisit some tools we've previously explored, but for now, let's look at what students can do with paper folding, basic manipulatives, and video.

Any time we make math hands-on, we tap into multiple senses and make learning more interactive (Lange, 2021). You're probably familiar with the concrete-representational-abstract (C-R-A) model popularized in the 1980s, but there are some misconceptions about it that can be damaging. This theory was adapted from Bruner's "enactive-iconic-symbolic" model, dating back to the 1960s, so it's not new. While there are decades of research extolling the benefits of concrete objects and drawings in understanding and representing math ideas, manipulatives are often perceived as "needed for only some students," "babyish," or only for "certain concepts." It is presumed that students who struggle benefit from this medium, but since all humans learn by "doing," they are great for *all* students of all ages (Leong et al., 2015). Therefore, in a Math Workshop Plus classroom, physical objects should be available to all students, at all times.

>
> **Connections to SEL**
>
> Viewing Behavior as Communication
>
> Have you ever tried to explain something and couldn't find the words? Can you remember a time you tried to describe something to someone who said they couldn't picture it and needed it *see* it? That's what it feels like for students who can't communicate effectively or express what they know orally or in writing. This is frustrating, confusing, and disheartening for students who may understand but can't show it. If you see students struggling with self-management during tests, the debrief, or the launch, offer the option of drawing a picture, modeling with manipulatives, or showing their thinking another way.

Do you have traditional manipulatives like base-ten blocks or 2-color counters? How about fraction circles, geoboards, or pattern blocks? Do you have students? What if *they* were the manipulative? How can students, as able, use their bodies as manipulatives? This is one (free) tool that you'll always have! In Chapter 6, we touched on students using their bodies to gesture, but that's just the beginning. From fraction operations and rounding to percentages and integers, so many ideas can be modeled this way! Could we assess this way? If students can communicate understanding without picking up a pencil or a device, and you can *see* it—meaning no correcting papers—it's a win-win! That meets our efficient, effective, easy, and equitable criteria!

Students can show their understanding of numbers, fractions, slope, angles, and more. How about multiplying two-digit by one-digit numbers? If able, students can use their bodies to decompose the two-digit number into tens and ones and physically model the distributive property. Any student who can use their arms freely can participate because standing is not required. If they *can* stand, this doubles as a movement break!

If you're intrigued but can't picture it or aren't sure how to do it, worry not because in true UDL form, we have videos! Access the videos using the QR codes found in Exhibit 9.1. As you watch, notice there are students sitting and standing because the task is accessible either way. See if you spot additional UDL elements shared in prior chapters. Get ready to get excited about what options you *can* offer students, regardless of your budget!

Exhibit 9.1 • Using Our Bodies to Communicate Understanding

Mingle & Count	Fraction Aerobics	Place Value Aerobics
qrs.ly/m1gn0f0	qrs.ly/legn0f6	qrs.ly/wagn0fa

To read a QR code, you must have a smartphone or tablet with a camera. We recommend that you download a QR code reader app that is made specifically for your phone or tablet brand.

What did you think? Are you ready to get started? Did seeing the videos spark ideas for other concepts? We like having students walk and jump on a number path or number line and act out number bonds. Think of the mileage you can get out of Fraction and Place Value Aerobics! How about decimal, percentage, and fraction equivalents, integers, or making a human coordinate grid to see the slope of a line?

We've included a few tips for additional supports to ensure these activities are as accessible as possible, even if there are physical or language barriers.

1. Post a visual of the "moves" (see examples in Figure 9.1). Think of it as a decoder key for students to use while their brains process the math.

Figure 9.1 • Fraction Aerobics

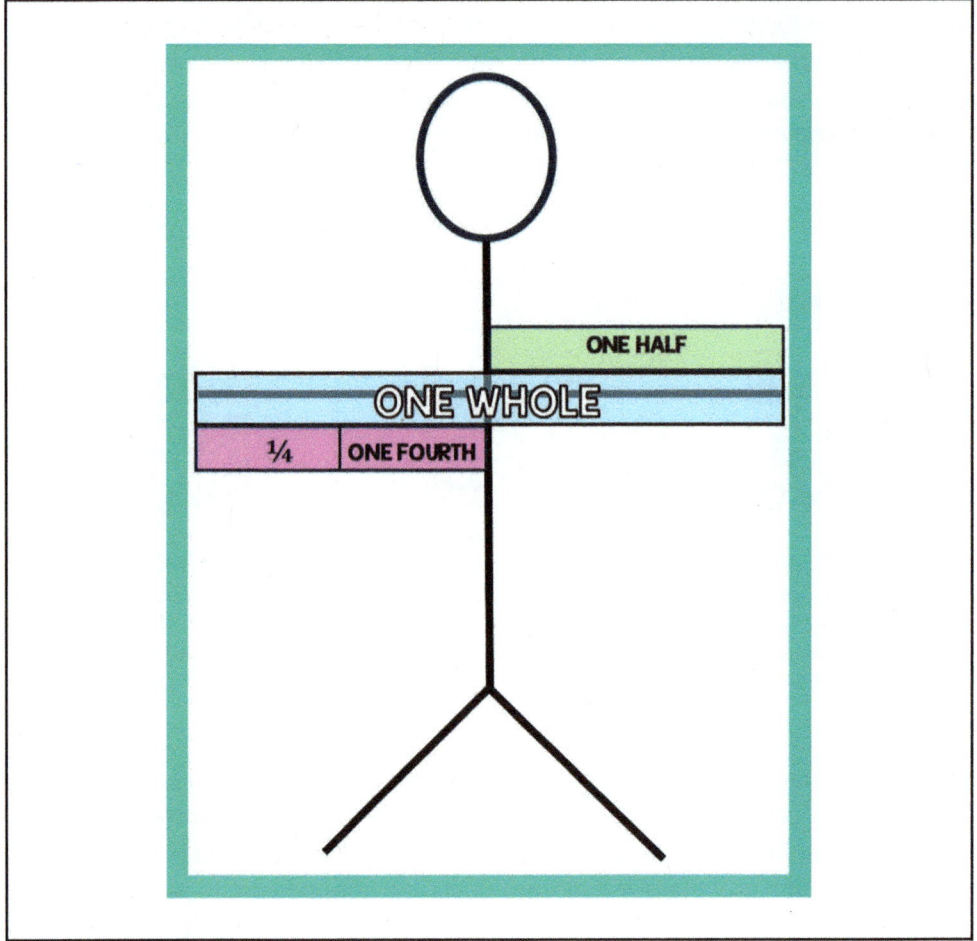

2. Post visual language supports. Create slides or record numbers that students are working with for place value aerobics on the board. Label the sides of the body as "tens" and "ones" and display a place value chart to add structure and clarity.

3. Repeat oral prompts multiple times. If possible, incorporate slides or write on the board. Remember to provide wait time.

4. For those with limited mobility, consider offering a toy to "act out" the moves from a chair or drawing them on a whiteboard.

5. At first, allow students to collaborate about how to model solutions. If you want to use it as an assessment, transition to having students model solutions independently.

Maybe you're excited, or maybe you can't really see yourself using these types of routines. Either way, give them a try. While this may not feel like your thing, it could be *the* thing that unlocks math for a student. Alison has done these activities with countless students and adults. In every case, connections are made and ah-ha moments abound while thinking, talking, and smiling!

Shifting back to common tools, we must mention calculators. Whether standard or graphing, calculators can be a handy tool when used sparingly and for a specific purpose. For example, say the goal is for students to demonstrate understanding of finding the volume of a cylinder. The dimensions include decimals, and decimal operations are a known barrier. Using a calculator for the decimals removes that barrier so we can see if the student understands volume. Do we care that they aren't proficient with decimals? Of course we do! That's why we love workstations and Guided Math groups. Both offer students opportunities to build proficiency as they go rather than pushing them along with gaps.

In recent years, equation editors have become more common, but a newer tool that is already a game-changer is artificial intelligence (AI). As AI in education evolves, it's worth noting that exposing students to these tools prepares them for the future. This technology responds to individuals, adapts to different communication styles, and offers personalized support in ways we cannot. While it may seem daunting, AI could be your secret weapon.

Express Learning in Flexible Ways

The idea that students can only show what they know through written assessments is as limiting as telling them that there is only one way to solve a problem! You can probably recall an instance where a student surprised you with their thinking when expressing it outside of a traditional test. Many state assessments accept visual representations, and there are additional options that may not only be more accessible but more engaging and effective for students—and more efficient for you!

◇◇◇◇◇◇◇◇◇◇◇◇◇◇◇◇◇◇◇◇◇◇◇◇◇◇

The idea that students can only show what they know through written assessments is as limiting as telling them that there is only one way to solve a problem!

Does formal assessment have to be done in a whole group? Consider the benefits of assessing in small groups. In small groups, students can use manipulatives to model answers with or without an oral or written response. It's unrealistic to suggest this for all students, but it's a helpful option when you suspect a student knows more than the traditional assessment shows. If small groups won't work, try technology. Rather than explaining their thinking to you, students record themselves. This ensures adequate time and offers you the freedom to watch and listen when time permits. One teacher shared how she uses the Seesaw student journal for this and how she watches over her morning coffee. She also loves how parents can view the recordings and says her parent–teacher conferences have improved as a result.

If this sounds too out-of-the-box for you, think of a time when a student bombed an assessment when you were sure they understood. That's an example of when to give it a try. We've heard stories from teachers who were stunned to see students who performed poorly on an assessment answer correctly when questioned orally. It's worth a try. What do you have to lose?

Use Multiple Tools for Construction, Composition, and Creativity

In Chapters 6 through 8, we explored some virtual manipulatives. We feel fortunate to be teaching in a time where so many rich, interactive resources are widely available. Digital media pairs well with this dimension of UDL because of its power and adaptability. We'll revisit some options, but before we do, let's check out a no-tech, old-school staple that's great for students to use for construction, composition, and creativity. We're referring to paper folding, and it can be powerful!

Over the years, we've had our share of debacles with paper folding, but when we look at it through a UDL lens, we see those challenges could've been minimized with more intentional thinking about potential barriers before we started. So before you decide to ditch paper folding forever, put your UDL hat on and remember the importance of creating avenues for students to construct ideas conceptually in your Math Workshop Plus classroom.

First, let's acknowledge that today's students are likely less experienced with paper folding. When we were kids, paper airplanes and origami were common. Maybe you can remember making folded-paper fortune tellers. If you don't know what we are referring to, Google *folded-paper fortune teller* and have some throwback fun! We remember folding the notes we were passing to our friends into neat squares and then tucking the corners in to make triangles. Looking back, that was super mathy!

Today, things are radically different. Outside of school, students have video games, tablets, and phones. In school, they can access technology, plastic or foam manipulatives, and other media. Therefore, there isn't a much of a need to fold paper. Back in the day, we did all sorts of cool stuff, and students got so much out of it because they were constructing ideas themselves.

When you try it, enter with the expectation that most students won't have experience with it and plan accordingly. Keep in mind that you should look at paper folding from an occupational therapy perspective, an executive-function perspective, and a spatial-reasoning perspective so that you can anticipate barriers. As you plan, flip your perspective from one that sees multiple barriers to one that is eager to capitalize on multiple opportunities. Through this lens, you will be more likely to recognize how many areas can be strengthened for students using this simple medium. You can also rely on good old paper if the power goes out or when the internet is down!

Here are our best tips for success:

- Be overly explicit when sharing instructions.
- If you have a document camera, use it!
- Provide visual examples for each step to support students who have limited language, weak executive-function skills, or struggle to stay regulated when things are tricky.
- Always have extra sheets of paper on standby!
- Try it in Guided Math groups instead of the whole group.
- Offer a paper-folding workstation so students can practice at their own pace and don't have the pressure of keeping up with you or with the person next to them.
  - Support this station with photos of the paper at each stage of folding and with visual cues, such as color-coded arrows to help guide students as they go.
  - Another option is to record a video students can access at the station. Since time is always at a premium, think about allowing someone else to "star" in the video. This could be a student, paraprofessional, or an older student looking for service hours. The video should only show the paper and the hands folding it, not faces.

Make mats available that are either color-coded or labeled to support students as they follow along with the video. These mats are also helpful if you engage in paper folding as a whole-group activity. See Figures 9.2, 9.3, 9.4, and 9.5 for examples of these mats. Note that the mats should be larger than the paper being folded. Note that the mat in Figure 9.5 is designed to reinforce geometric vocabulary and assumes you have visual vocabulary on display.

Figure 9.2 • Color-Coded Mat to Support Paper-Folding Practice

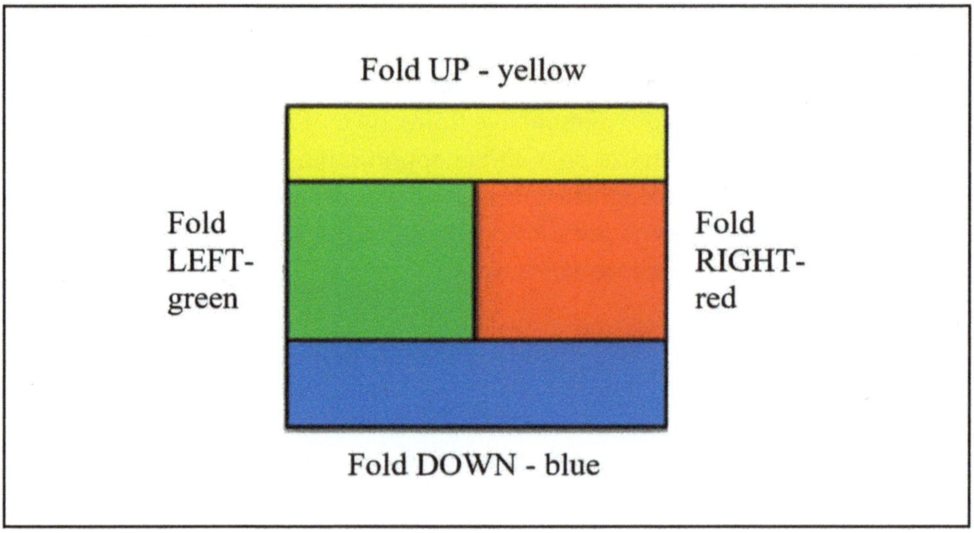

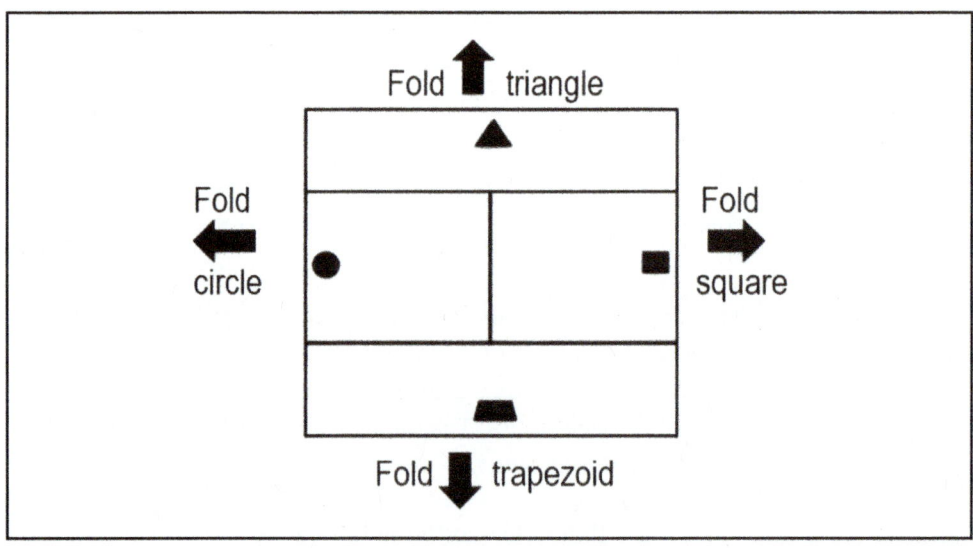

Figure 9.3 • Symbol Mat With Shapes to Support Paper-Folding Practice

Figure 9.4 • Symbol Mat With Emojis to Support Paper-Folding Practice

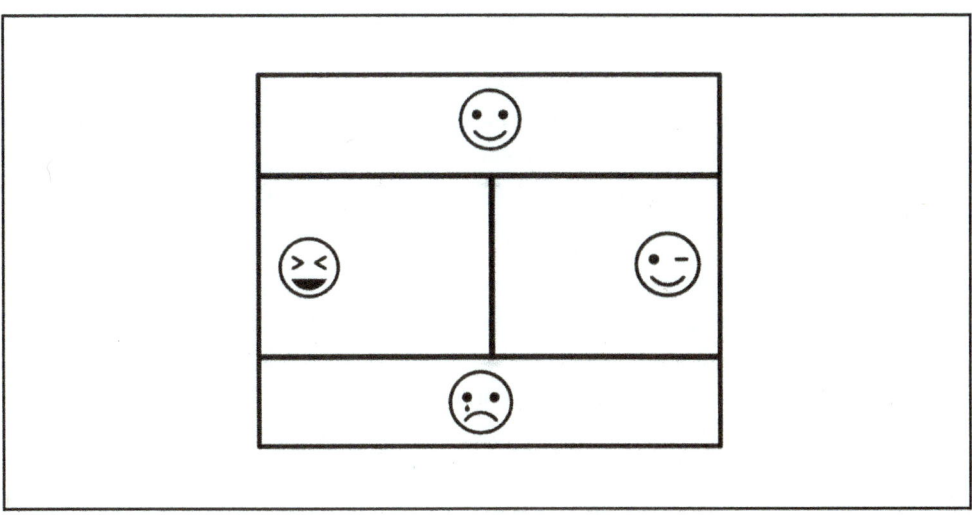

Figure 9.5 • Mat With Model Fold Lines to Support Paper-Folding Practice

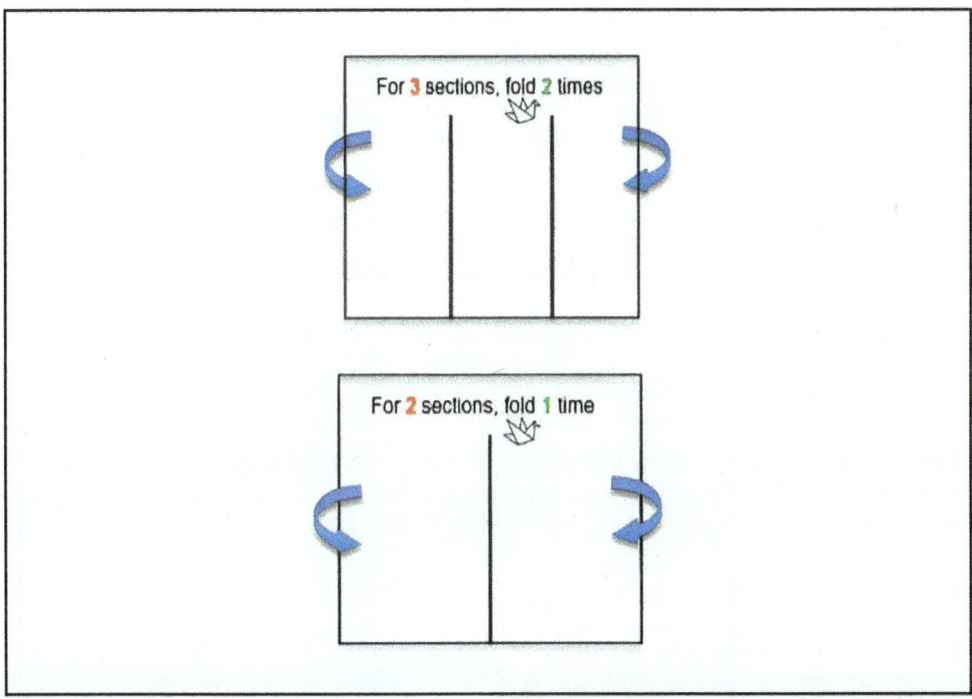

> **Equity Check**
>
>
>
> Monitoring Your Analogies
>
> We did not include folding the "hamburger" or "hotdog" ways or the "garage door" or "front door" ways. We used these back in the day, but now we want to minimize language and be more intentional about equity. Not all students know these words, what they mean, or what we are trying to convey when we use them as analogies.

Let's explore what concepts paper folding works for and if you'll use it during mini lessons, Guided Math groups, or workstations. The ideas in Exhibit 9.2 (Figures 9.6–9.11) may not be new or groundbreaking, but perhaps they will remind you of something you have forgotten or encourage you to revisit something that may be more successful when you try it again with your UDL hat on. Either way, we hope that these ideas reinforce the point that you don't have to have fancy manipulatives to provide additional options for students to show what they know.

Exhibit 9.2 • Simple Ways of Showing Concepts on Paper

Fraction Comparison	Equivalent Fractions	Fraction Multiplication
Figure 9.6 • Comparing Fractions	**Figure 9.7 • Equivalent Fractions**	**Figure 9.8 • Multiplying Fractions**

Multiplication: Arrays and Area Models	Area of a Triangle	Symmetry: Create Examples and Nonexamples
Figure 9.9 • Area Model of Multiplication	**Figure 9.10 • Discovering the Formula for Area of a Triangle**	**Figure 9.11 • Symmetry**

We have already made it super clear that concrete models are our go-to so students can experience all stages of the C-R-A progression as they learn mathematics. We've seen the power of building area models of multiplication with base-ten blocks before drawing, using the beaded number line in conjunction with the written number line, and discovering integer rules with 2-color chips or algebra tiles instead of memorized rules. While physical manipulatives are our first choice, not every school, classroom, or teacher has access to them. We regularly hear from special educators that they don't have access to the same resources as general educators, something that is extremely frustrating, disappointing, and counterintuitive as far as we are concerned. Luckily, there are alternatives for anyone with a computer, tablet, or interactive board.

If you have one of these devices and are eager to use manipulatives but don't have access to them, get ready to go virtual! As previously mentioned, there are many free virtual manipulatives out there, and although they are semiconcrete, they do offer a language-free visual that can be *manipulated*, so let's think about how you can use them. With most students, we use them *in addition to*, rather than *instead of*, physical manipulatives. For some students, however, virtual manipulatives may be even better than physical manipulatives. If you've been teaching for more than a couple of years, we're pretty sure you can think of a bunch of reasons for this, but just in case, we outlined a few for you.

Virtual manipulatives are great because of what they *don't* do. They don't (usually) cost money. They don't fall on the floor. They don't make noise. They don't get lost. They don't require students to get them, or teachers to pass them out. They don't get dirty. They don't wear out. They don't have fixed sizes or colors, and you can access as many or as few as you need in an instant. There are countless ways they can be customized, organized, and personalized. In other words, they are a perfect match for Math Workshop Plus!

Let's look at a math example from each level and imagine how these digital tools for construction and composition can be great options for students to communicate their understanding in relation to the learning goal. We'll peek into a first-grade classroom (Exhibit 9.3, Figures 9.12–9.14) to see how students use an interactive hundreds chart and digital number line to show their understanding of 10 more/10 less. As you explore the examples, think of what the different tools offer and how they complement the learning target.

Exhibit 9.3 • First-Grade Example of Digital Tools

Learning Target: Students will communicate an understanding of 10 more/10 less using an interactive hundreds chart and interactive digital number line.

Part I:

1. The starting number is marked yellow by the teacher or the student.
2. The student identifies the number that is 10 more with green.
3. The student identifies the number that is 10 less with red.
4. Student repeats several times and points to or explains which digit is changing each time.

Figure 9.12 • An Interactive Number Chart

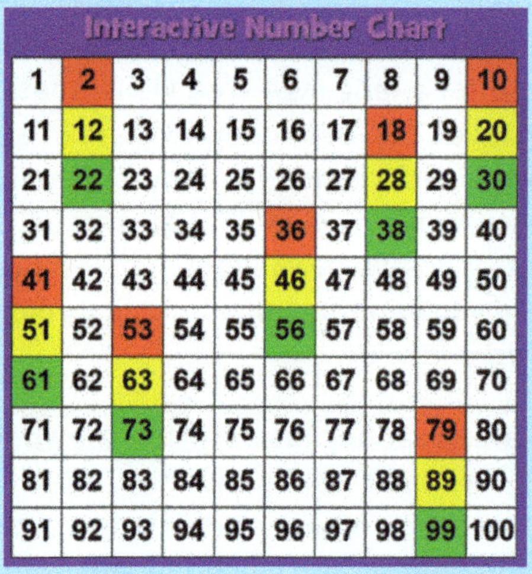

Source: Math Playground

Part II:

1. The teacher shows a card with a starting number.
2. The student identifies that number in yellow on the number line.
3. The student rolls a cube labeled +10/−10.
4. The student models the solution on the number line using a green jump forward for 10 more, and a red jump back for 10 less.

5. The student labels the ending number with green (10 more) or red (10 less).

6. Student repeats several times and points to or explains which digit is changing each time.

Figure 9.13 • A Number Line Illustrating a Jump of 10 Forward to Add

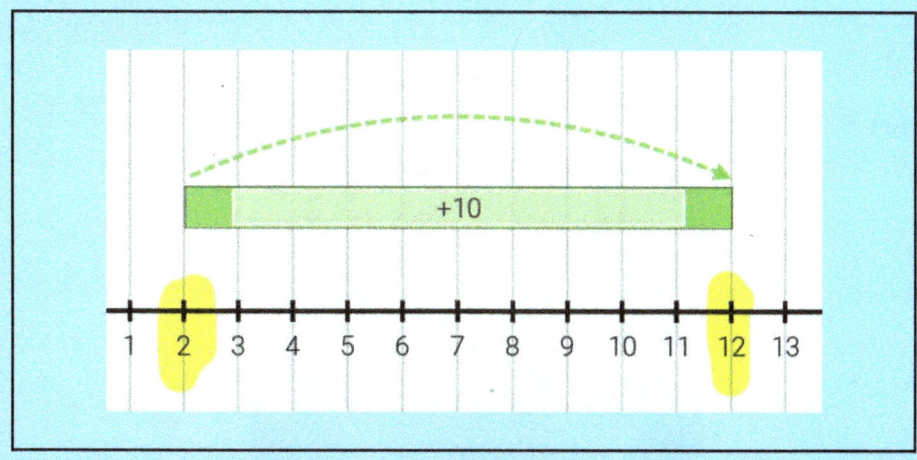

Figure 9.14 • A Number Line Illustrating a Jump of 10 Backward to Subtract

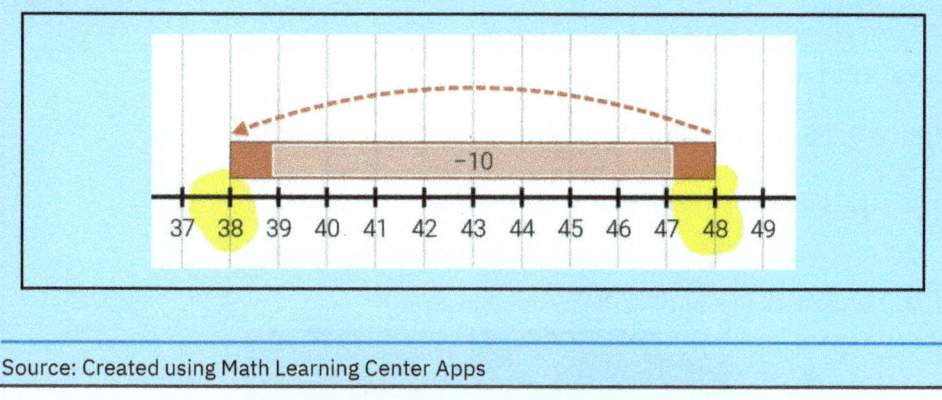

Source: Created using Math Learning Center Apps

Exhibit 9.4 shows how students in Grade 4 discover the relationship between equivalent fractions with digital pattern blocks and virtual fraction bars (Figures 9.15 and 9.16).

Chapter 9 • Design Options for Expression and Communication **241**

Exhibit 9.4 • Fourth-Grade Example of Digital Tools

Learning Target: Students will communicate an understanding of equivalent fractions using digital pattern blocks or digital fraction bars.

1. Establish which block represents one whole.
2. Use other pattern blocks to build equivalent fractions.

In this example, it is clear that the red trapezoid is $\frac{1}{2}$ of the hexagon and the green triangles also represent $\frac{1}{2}$ but is represented as $\frac{3}{6}$.

Figure 9.15 • Pattern Blocks Modeling Equivalent Fractions

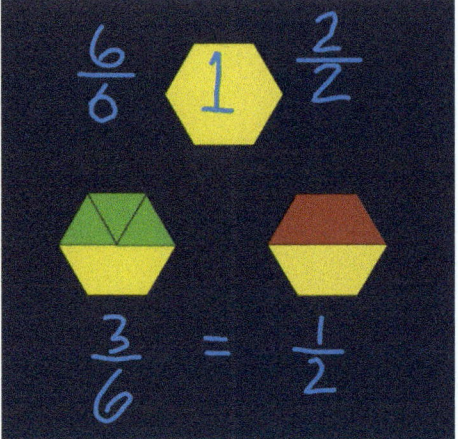

Source: Math Learning Center Apps

Figure 9.16 • Fraction Bars Modeling Equivalent Fractions

1. Use the fraction bar tool.
2. Drag fractions of different sizes to see where they match up.
3. Note those that align perfectly as equivalent.

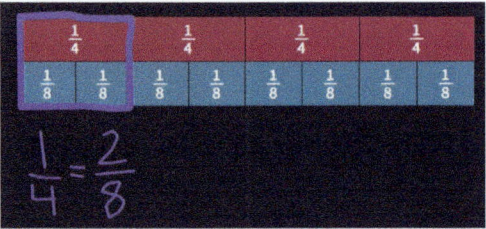

Source: *Materials created using Desmos Classroom, an Amplify product. Visit Desmos Classroom for lessons, lesson building tools, and Polypad virtual manipulatives at teacher.desmos.com. Corwin is not affiliated, sponsored, or endorsed by Amplify.*

Finally, let's look at a Grade 7 example (Exhibit 9.5, Figure 9.17) and see how students can explain how to solve surface-area problems using a digital tool that allows them to unfold and decompose 3-dimensional shapes to see the 2-dimensional shapes they are composed of, helping them to make sense of the pieces they are finding the areas of.

Exhibit 9.5 • Seventh-Grade Example of Digital Tools

Learning Target: Students will communicate an understanding of solving surface-area problems using digital 3D figures that unfold to reveal the polygons they are composed of and combining their areas.

1. Open the 3D figure tool in Polypad.

2. Use the "unfold" feature to reveal the polygons that make up the figure.

3. Note the number of "copies" of each polygon (in Figure 9.17, six rectangles and two hexagons).

Figure 9.17 • Virtual Hexagonal Prism Unfolded to Explore Surface Area

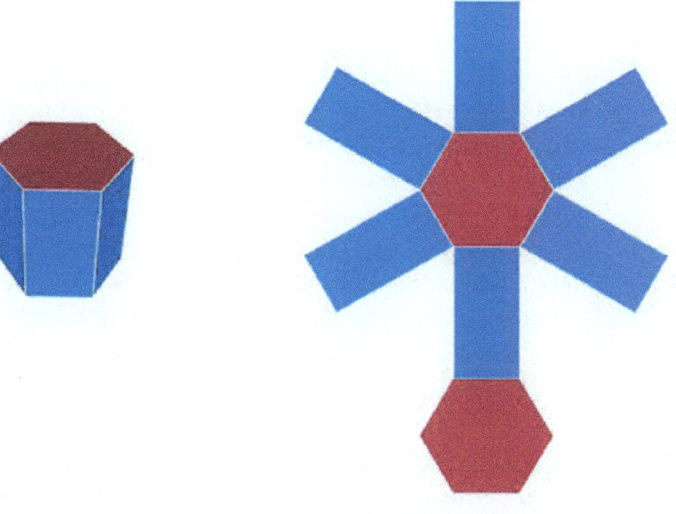

Credit: Materials created using Desmos Classroom, an Amplify product.

Connections to SEL

Avoiding Assumptions That May Result in Frustration

Even though your students are digital natives (Prensky, 2001), that doesn't guarantee their skills are a match for what is needed in the classroom. They may be great at gaming and swiping, but not at understanding how the space allotted for an open-response question adjusts as you type or how to use a digital ruler. For students to use these virtual tools effectively, you must explicitly teach them how. When you do, you empower them to take the initiative to use what they need, believe they are capable, solve their own problems, and successfully demonstrate their learning and communicate their thinking. If you don't, these experiences are at best a waste of time and, at worst, work against your efforts to promote a growth mindset, self-efficacy, and self-management.

Equity Check

Making the Most of What You *Do* Have

During the COVID-19 pandemic, many schools and districts ensured that every student had a device to use. Unfortunately, not all schools had the ability to do this, and many who *did* didn't have the ability to sustain them. The result was, and may still be, a lack of access for some students. If that's the case, you can create a similar experience in your classroom by allowing students to use a document camera and physical manipulatives or to access the interactive board or overhead projector while you meet with Guided Math groups. While we cannot ensure that all districts, schools, and classrooms are provisioned in equitable ways, we can help you more efficiently and effectively use the resources you do have to create avenues for all students to express and communicate what they know and are able to do.

Build Fluencies With Graduated Support for Practice and Performance

This section is one that we have been waiting to get to because there is so much to unpack. It offers a chance to ensure that as we build Math Workshop Plus and think intentionally about UDL, SEL, and equity, we also highlight what best practices in math instruction look like, sound like, and feel like for students. In this section, we'll explore this element of the UDL framework and what specific teacher moves you want to integrate, amplify, and/or practice. Maybe you'll also identify some practices to leave behind, so get ready with your **x**, ★, ⬭ and ! Although this book is centered around barrier removal and support for students, we want to differentiate between *effective* supports that foster independence and productive struggle and *ineffective* supports that create student dependence and can result in learned helplessness.

Apply and Gradually Release Scaffolds to Support Independent Learning

As the title of this section says, you want to generate and maximize opportunities to build fluencies with graduated levels of support for practice, and Math Workshop Plus offers a great structure to do this, but before we jump into some ideas and examples, think about how you define *support*. It is important to pause here and answer that question because your response could be the difference between support that fosters independence or support that does the opposite. Once you have your definition, review Table 9.1 and assess where your current approach falls so you can uncover if it is working *for you* or *against you* in reaching the ultimate goal of learner agency.

Table 9.1 • Supports That Foster Independence

Action to Avoid	Because	A Better Support
Going around to check student work or having students come to you to have it checked	Students become accustomed to needing validation from you before moving on.	Make work self-checking by posting answer keys or using QR codes.* *Note*: We know the value of reviewing student work and suggest students note which problems were incorrect and hand in the work so you can review it. This can be great formative assessment to form Guided Math groups.

(Continued)

(Continued)

Action to Avoid	Because	A Better Support
Allowing students to interrupt your Guided Math group with their questions	Students need to learn how to utilize the resources available to help themselves answer their own questions.	Create the "What to Do if I'm Stuck" anchor chart seen in Chapter 4 and do not engage when students attempt to interrupt you. Instead, point to the chart. If they remain stuck after trying a tool, checking their Reference Folder, and asking some classmates, they may write their name on a designated spot on the board and move on to a different workstation. When you wrap up your group, you can check in and see what the student needs.
Telling students which tools to use or controlling the tools (i.e., passing them out for a specific lesson rather than having them available at all times)	Students will wait until you direct them. They will not learn how to match the tool with the task. They also may not discover which tools work best for them.	Create a "tool buffet" or individual tool kits so that students may access tools such as manipulatives, rulers, paper for folding, and so on whenever they want to use them.
Using manipulatives in Guided Math groups only when students are struggling with concepts	Students need to know that manipulatives may be used at any time and not just by "some students."	Make a practice of connecting the concrete to the representational and the abstract simultaneously so that students see and understand the connection between them. This is especially important when working with concrete models so students have an opportunity to transfer what they are "doing" to the abstract representation since that is what they'll see when working independently.

Action to Avoid	Because	A Better Support
Calling students to the teacher table to complete the Must Do or independent practice	Students will expect to be guided through the work that should be independent. Although well-intentioned, impact trumps intent here, and the impact is that students become trained *not* to try their work independently.	Be sure students have what they need to be successful. If you want to help without helping, do the first one together before releasing students or ask questions such as these: • What do you know? • Where will you start? • What is being asked? • What tools might you use to help you? • Do you plan to work alone or with a partner?

When students learn about how they learn, they become strategic and action-oriented to support themselves. They learn what they need to be successful and access it until they no longer need it. As the guide on the side, you can ask curious questions and encourage students to discard scaffolds as their learning progresses.

We love the idea of contrasting "just-in-time" scaffolds with "just in case" scaffolds (Dixon et al., 2019). Just-in-time scaffolds are more personalized and can be adjusted by the student or by the teacher. These scaffolds are used because they are needed to remove a barrier, not just because they exist and therefore should be used. Such scaffolds allow students to think, reason, and decide what is needed, rather than relying on an adult to break it down into tiny bites for them. Dixon and her team (2019) refer to these as tools that develop "productive perseverance" because they help students remain engaged even when problems are challenging. A great example of what this looks like is often seen when students are working on an adaptive, computer-based program for extra practice. As they progress, the computer adjusts to their responses and serves up questions that are designed to be in the Zone of Proximal Development (ZPD) for that student (Vygotsky, 1978). If, however, the student gets stuck or answers incorrectly, they may see a pop-up that asks if they wish to view a hint. The hint is a just-in-time scaffold. It wasn't there until the student faced a barrier because its purpose is to remove the barrier.

Connections to SEL

Scaffolding to Support SEL

Students with lagging SEL skills may need more scaffolding to engage in "productive perseverance." For these students, having checklists to keep themselves calm, know what steps to take if they get stuck, and guide them to work collaboratively with others can make a big difference. These students may need more frequent reminders that mistakes are part of learning and that speed is not an indicator of aptitude. These reminders will support growth mindset, build confidence, and reinforce the idea of self-efficacy.

In a Math Workshop Plus classroom, students have frequent opportunities to practice their math fluency. Whether we are thinking about computational fluency, procedural fluency, or fluency with basic math facts, Math Workshop Plus offers space in every aspect of the math block to practice with graduated levels of support. The workshop might begin with a Number Talk, which is designed to give students frequent opportunities to play with, think about, and discuss numbers. Students share strategies, listen to one another, and look for efficient ways to solve problems. While there are many benefits to this routine, ranging from risk-taking to positive and supportive culture and improved number sense, the main goal is to develop computational fluency supported by conceptual understanding.

The mini lesson can offer another opportunity to practice procedural fluency. If your instruction is grounded in best practices in mathematics, students will build flexibility with numbers and shapes, carry out procedures with accuracy, and strive for efficiency. Following the mini lesson, students should have a chance to continue to practice independently. As we just discussed, for students to be successful with independent practice, we must allow them to be independent!

When you call students to meet in a Guided Math group, they could be working on either computational fluency, procedural fluency, or even basic fact fluency. The small-group setting is perfect to differentiate scaffolds and ensure students are choosing what they need to be successful. We expect students to join us in Guided Math groups with their Reference Folders in hand and their individual tool kits (if they have them). In this way, students are in charge of adjusting their scaffolds and learning about themselves as learners. Of course, if you notice students aren't using their tools and should be—or that they're using tools they don't need—use questioning to help them better understand where they are and what they need.

Most people who use Math Workshop carve out a consistent workstation that is focused on fluency with basic facts. Depending on the grade, students may opt to use a Rekenrek, fingers, a beaded number line, area model, open number line, or some other tool to support their reasoning. They may also choose to use their Reference Folder or visual flashcards. Whatever the case may be, students can apply and release the right scaffolds for them according to their readiness and confidence.

Connections to SEL

Pay close attention to students who aren't making the expected progress with basic fact fluency. Use questioning to reveal their mindset. If they are focused on speed alone, they may be unwilling to view hints, utilize tools, or reason through a problem. If students become self-conscious, anxious, or frustrated, they may struggle with self-management. This may manifest in the form of behaviors and emotions such as anger and sadness, self-doubt, and a fixed-mindset. If a student needs scaffolds to be successful, ask some questions to get them thinking and moving in the right direction. Remind them that they can set small, manageable goals, and be sure to celebrate even small gains so that they can focus on their progress and build their self-efficacy.

Address Biases Related to Modes of Expression and Communication

Using a variety of modalities and media to communicate should be celebrated, but not everyone may see it that way. Students should be comfortable using what they need without fear of judgement or ridicule. To foster this culture, normalize multiple means of action and expression; educate your students, colleagues, and anyone else who interacts with your students.

Students should be comfortable using what they need without fear of judgement or ridicule.

As previously noted, language frames are recommended to support communication. If you regularly engage in routines and number talks and expect all students to share their strategies, normalizing access to sentence stems, word banks, visual vocabulary, and speech-to-text devices communicates to students that they have options to support their success and that you value their contributions regardless of the modality in which they are expressed. In a Guided Math group, this may look like a tented sentence stem between two students, with one sentence for Student A and the other for Student B. These are also helpful during the launch or mini lesson to support turn-and-talk opportunities. This may also be accomplished with speech to text or passing a translation device between two students or between the student(s) and you.

> **Say Bye-Bye to Barriers When You . . .**
> - Draft your "start," "stop," and "continue" list
> - Encourage communication through various media
> - Make manipulatives available on demand
> - Normalize using tools to show understanding
> - Have students use their bodies as a math tool
> - Try paper folding
> - Provide just-in-time scaffolds
> - Offer virtual manipulatives

Action Plan

Assess your current options for Design Options for Expression & Communication. Download the Chapter 9 Self-Assessment at https://companion.corwin.com/courses/MathWorkshopPlus.

Try It! Adapting and Modifying Routines

View the videos of fraction and place-value aerobics. Try the routine with your students and note any adaptations or modifications you make.

Connections

We're confident that you will run away with ideas in this chapter and use them with concepts we haven't even considered. When you do, please share on social media with #mathworkshopPLUS so *you* can teach *us*!

◇◇◇◇◇◇◇◇◇◇◇◇◇◇◇◇◇◇◇◇
Chapter 10

Design Options for Strategy Development

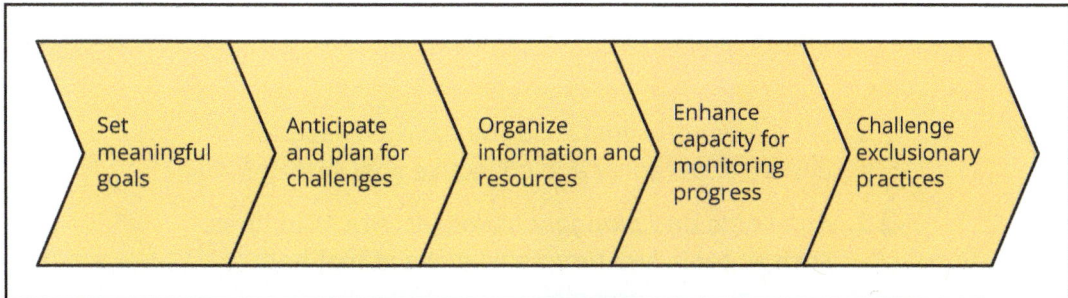

As you near the end of this book and envision your classroom through the lens of Math Workshop Plus, you may feel apprehensive about the level of autonomy students will have. Perhaps you are thinking that they won't be able to handle it, or you wonder if they really will stay on task and use their minutes wisely. It's normal to feel that way. It's scary to feel like you're giving away control, but remind yourself that you're not just throwing students in and wishing them luck! Remember that you will prepare students for this and set them up for success. We've mentioned the First 20 Days countless times because what you during that time is super important! As you establish systems to maintain accountability, co-create anchor charts, and develop and maintain routines and expectations, you curate opportunities for students to hone their executive-function skills. When you make it clear that you expect them to focus on their work, pay attention, work independently, work with each other, wait for each other, and take turns, you set the stage for authentic practice. You position students to set goals and work to achieve them. You meet with them to strategize how to reach their goals and how to reflect on and revise their work in pursuit of those goals. This is an important life skill that will serve them well beyond math! Research states that children are "not born with these skills but they are developed over time" (Harvard University, n.d.).

As the teacher leading Math Workshop Plus, you design the learning environment that fosters the development of these skills, which gives students the priceless gift of learning how they learn.

> "Executive function skills have been shown in many research studies to be related to math achievement. Children with better executive function skills including but not limited to attention, working memory and cognitive flexibility succeed in math with greater ease than their less skilled peers" (Kenney, 2022, para. 10).

To successfully create a learning environment where students are enthusiastic, focused, knowledgeable about themselves as learners, and assessment savvy, ask yourself these questions about the five facets of the *Design Options for Strategy Development*:

1. *Set meaningful goals*: Do you have a classroom culture of personal and collective goal setting?

2. *Anticipate and plan for challenges*: Are structures such as tools and templates in place that help students to be organized, thoughtful, and independent in their learning journey?

3. *Organize information and resources*: How are connections made, and how are information and resources organized so that demands on working memory are minimized?

4. *Enhance capacity for monitoring progress*: What systems are in place for students to receive feedback and monitor, reflect, and evaluate their work throughout units of study?

5. *Challenge exclusionary practices*: What shifts might need to be made to build and maintain an inclusive classroom that "names, explores and addresses" bias and exclusionary practices?

Set Meaningful Goals

Research shows that goal setting can be a game changer in terms of student achievement (Midwest Comprehensive Center, 2018). When students are focused, they know what they are working toward and what it looks like when they achieve it. When they track their progress along the way and celebrate milestones, they are more motivated and stay engaged. Math Workshop Plus creates a student-centered, personalized environment that is driven by student goals. You model this when you share the goals of your mini lesson and when you discuss the learning goals and success criteria of each of your Guided Math groups. Workstations represent purposeful practice opportunities for targeted skills and concepts, so to ensure they align with goals, they must be clearly marked with the learning targets. This helps to ensure that students make just right choices that support what they are working toward. During the debrief, you can ask students what milestones they may have hit and how they are feeling about their progress.

While you will offer significant input and recommendations to students, you should not set goals for them. Instead, you will designate time for goal-planning sessions with students to determine where they are and what they need to do. By allowing students to set their goals, they will own them. When they own them, they are intrinsically motivated to meet them. Keep in mind that goal setting is a skill that you will have to explicitly teach just as you would any other skill. Taking time in the beginning of the year to describe what it is, why you do it, and to model it so students understand what the criteria for success looks like are all critical before having students do it.

The goals that students set should be both "challenging and realistic" (CAST, 2024). Offering scaffolded guides and checklists along the way helps students to continually monitor their goals. CAST advocates for posting goals, objectives, and schedules publicly. We embrace publicly posting *class* goals and progress as a way to motivate and celebrate, but we cringe when we see *individual* data or progress toward goals posted because that is private information. This should never be made public for others to see and comment on because, within the community of learners, students are individuals, and their journeys are unique to them. We keep student data in folders, where students can revisit and reflect on their individual data with you and use it adjust goals, celebrate success, and set new goals.

Use Prompts and Scaffolds to Estimate Effort, Resources, and Challenges

Goal-planning templates are a great way to get started with goal setting. Their structure creates a space for students to build awareness of themselves as learners. With these organizers, students assess where they are, where they need to go (informed by some guidance from you), and what steps they need to take to get there. They can also reflect on their effort, how they use the resources available to them, and what level of challenge they are feeling. Notice that the action plan is embedded into the design (see Exhibit 10.1).

Exhibit 10.1 • Goal Planner

My Goal Planner

Goal: Learn My Double Facts

1. **By Myself**

   Study my facts, practice with my games and my visual flashcards

2. **With My Teacher**

   Work in the Guided Math group, play games, discuss and make up my own problems and flashcards

3. **With My Friends (Workstations)**

   Play a doubles game:
   - Cards
   - Dice
   - Dominos

4. **At Home**

   Play games at home, practice with my visual flashcards

   Resources I Will Use:
   - Rekenrek
   - Cubes
   - Strategy Cards

Challenge Level: (place an **X** on the line to show)

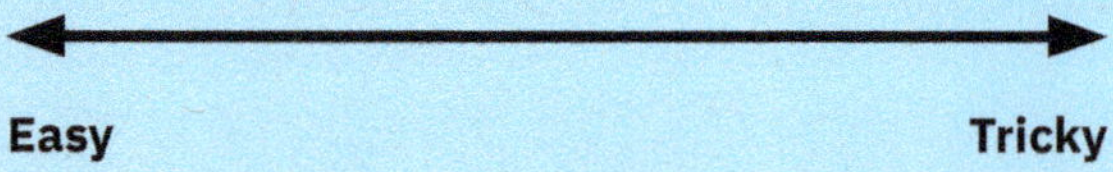

◄─────────────────────►

Easy   **Tricky**

Source: Newton, 2024.

Goals take time and intention to set, effort and action to pursue, and perseverance and patience to achieve. Students should think about these things as they are setting their goals. Attaching an action plan to their goals is essential to helping students connect their progress to their effort and reinforcing the idea that they can achieve their goals if they try their best and stick with it. We have seen some teachers end their week with a reflection called an "effort meter" (see Figures 10.1 and 10.2). An effort meter prompts students to engage in honest reflection, assess themselves, and share their assessment of their effort that week with a peer and/or you. We have found that they are super honest!

Figure 10.1 • Effort Meter Example 1

GREAT EFFORT
I did it. I stuck with it. I asked for help if I needed it. I persevered.
GOOD EFFORT
I tried. I stuck with it. There are some things I could do better.
SOME EFFORT
I started. I tried. I could have tried harder.
LITTLE EFFORT
I gave it a shot, but I didn't stick with it. I gave up right away.
NO EFFORT
I didn't even try.

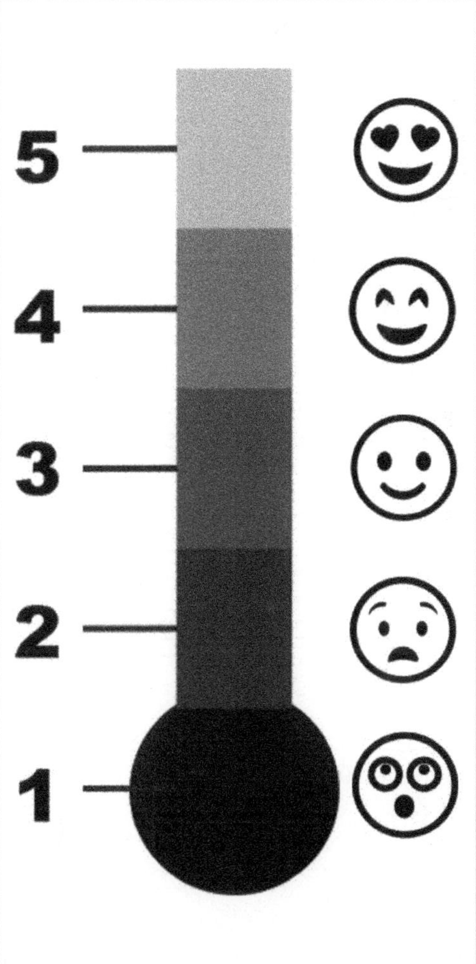

Emoji source: Freepik.com/jcomp

 Download this Effort Meter at https://companion.corwin.com/courses/MathWorkshopPlus.

Figure 10.2 • Effort Meter Example 2

MY EFFORT METER

How much effort did I give this week?

1 — No EFFORT
2 — Little EFFORT
3 — Some EFFORT
4 — Good EFFORT
5 — Great EFFORT

Record your rating each day at the end of Math Workshop.

Monday	
Tuesday	
Wednesday	
Thursday	
Friday	
Total	

HOW DO YOU FEEL ABOUT YOUR EFFORT THIS WEEK?

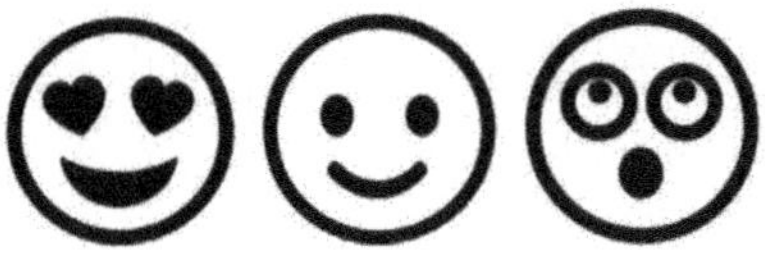

What adjustments will you make next week?

Source: Newton, 2024. Emoji source: Freepik.com/jcomp

 Download this Effort Meter at https://companion.corwin.com/courses/MathWorkshopPlus.

Connections to SEL

Self-Management and Self-Awareness

As students use various scaffolds to plan how they will accomplish their goals and monitor their progress, they are engaging in self-management. This is a skill that takes time to master. The process of creating opportunities for self-reflection and using tools like the effort meter helps students to think about how much effort they are personally putting into learning the topic, which builds self-awareness. Creating the conditions to build these competencies helps students see that the person most responsible for their learning is them, not the adults who support them.

Equity Check

Building Executive-Functioning Skills

Planning with diversity in mind is the foundation that Math Workshop Plus is built on. An equity-based approach to cultivating executive-function skills considers individual differences among students. Students need scaffolds for these critical non-math skills because they greatly affect progress and can get in the way of students accessing learning, functioning effectively in the learning environment, or showing what they are capable of. These may include planning ahead, meeting goals, organizing work, reflecting on goals and successes, displaying self-control, and staying focused (Center on the Developing Child, n.d.).

Anticipate and Plan for Challenges

One of your many roles as a teacher is to help students learn how to make a plan before they just jump into an assignment. By anticipating and planning for potential challenges, more students have opportunities to successfully navigate their learning. In your classroom, you know what tools, resources, and templates are available to scaffold the journey and help students navigate some of the trickier areas. Collaborating with students as they make their plan builds a script for their current and future selves and helps them become attuned to what they need and what works for them. As noted by CAST, it is important "to help learners become more planful and strategic" (2024). Scaffolds that help students stop and think about what they are doing, reflect on why they are doing it, and help determine whether to continue or change course when necessary build resilience and help students persevere.

Graphic organizers and templates are great scaffolds to support and sustain student thinking. Think about the things you saw in the Reference Folder, such as number lines, worked examples, and templates, then add scoring rubrics, checklists, and note-taking tools. These supports help students think about what they are doing and get it down on paper where they can see it and reflect on it in different ways. Whether you use the Reference Folder idea or not, these templates are a key to independence and support students in showing what they know in different ways. Be sure that they are easily accessible and promote their use by any and all students until they are no longer helpful.

Problem-Solving Templates

Problem-solving templates are organizers that help to guide thinking about solving problems, the strategies and models available, and how to reflect on the reasonableness of solutions. These templates help students to navigate problems in a systematic and explicit way. Instead of jumping in and grabbing numbers, students can think about the actual problem, develop a plan, and then reflect on it in a guided, scaffolded way. By having students check the problem in a different way than they originally solved it, they learn to double-check their work.

Problem-solving templates help students to "stop and think" and encourage them to "show and explain" their work (CAST, 2024). CAST notes that we should "provide checklists and project planning templates for understanding the problem, setting up prioritization, sequences, and schedules of steps." Templates, rubrics, and checklists all serve to coach students on their personal learning trajectory. If you currently use a template, compare and contrast it with Exhibit 10.2. Ask yourself if it promotes thinking and understanding or answer getting.

Exhibit 10.2 • Problem-Solving Template

Problem	How will you model it?
What's this problem about? Is it 1-step? Is it 2-step?	Number line Sketch Ten frames Hundred grid Other?
What is another way to check your answer? Number line Sketch Ten frames Hundred grid Other?	**Double-check your work.** Check the math _____ Check the answer _____
Answer: _____	Does this make sense?

Source: Newton, 2024.

 Download this Problem-Solving Template at https://companion.corwin.com/courses/MathWorkshopPlus.

Problem-Solving Checklist

Rubrics and checklists also prompt students to stop, think, and reflect on what they are doing. They offer helpful prompts and questions that push students to reflect on what they've done well and what might need to be revisited. Notice how this template shown in Table 10.1 helps students to reflect on both the "quality and completeness" of their work (CAST, 2024).

Table 10.1 • Problem-Solving Checklist

Add a Check to Each Item as You Complete It!	✓
Read the problem **two** times.	
Visualize the problem. (Make a picture in your head.)	
Translate the problem. (Say it or think about it using your own words.)	
Make a plan. What will you do to solve this?	
What models will you use? List one to solve and then another to check.	
What strategies will you use?	
Does your answer make sense?	
Did you double-check the work? Check your answer. Check your model.	

Source: Newton, 2024.

 Download this Problem-Solving Checklist at https://companion.corwin.com/courses/MathWorkshopPlus.

Templates and Visual Organizers: Tools for Seeing Information

Let's look at a middle school example. Exhibit 10.3 is a graphic organizer to help students think about proportional relationships. The information is organized in a way that supports connections and sense-making and includes visual elements rather than just text. The structure prompts students to look at, think about, and make connections about the concept.

Exhibit 10.3 • Proportional Relationships

Proportional Relationships
Equation:
Table / Graph
Story / Description

>  ### Connections to SEL
>
> #### Peer Work
>
> Working together on graphic organizers and ways to represent and evaluate their work is a great way for students to learn to communicate clearly, be helpful, and reflect not only on the learning of others but also on their own work. They learn to work together through cooperative structures, listen to each other, negotiate, and compromise in order to get the work completed. Students learn also how to manage their emotions and how to speak with each other. They also learn to show empathy for others. This is a great way to get students to build and maintain relationships with a variety of students in the classroom.

Equity Check

Planning For All

This entire book is designed to help you think about equity as you plan. Planning is about making sure every student in your class has the opportunity to access the standards. You will constantly be asking yourself, "Who needs what to be successful throughout the unit?" Your investment in intentional planning for equitable access will result in powerful learning for all students throughout Math Workshop Plus. In the past, your planning may not have considered the needs of all students in such depth. Think about this, especially if you're an experienced Math Workshopper who has struggled with "certain students." Shifting to planning through an equity lens will push you to ensure that, with the correct supports in place, every student is engaged, can access the content, and can accomplish the standards (Miller, 2022).

Organize Information and Resources

As students take more ownership of their learning, you may find yourself talking less, being more strategic about your questioning, and more thoughtfully orchestrating opportunities for students to make connections. Students come with a wealth of prior knowledge, even if they don't know it! So much of what you do in Math Workshop Plus is designed to help mobilize the skills, knowledge, and resources students have at their disposal, so their working memory can be available for new learning. When you make organizers and tools available, it helps reduce the cognitive load and creates fertile ground for new learning. The addition of these structures showcases their value as students experience the benefits of keeping information organized and the efficiency of using patterns and discovering themes.

Offering multiple ways to make connections increases the likelihood that students will connect with ideas and recall them later. By including real-life examples, math models, and explanations that they write, record, or demonstrate, students see the concept in different ways and are more likely to attach their new learning to something they already know. Having students discuss and explain these different representations with a partner takes it a step further. A graphic organizer provides space to record a variety of models, thoughts, and conceptions so that students can see the concepts in different ways, recall it in different ways, and maybe try a more efficient approach next time.

If you haven't previously used them, guided checklists could be your new secret weapon for students who are wrestling with executive-functioning skills. To maximize effectiveness, it's important to use them throughout the year and begin with activities that are heavily scaffolded before slowly fading out the scaffolds. In this way, students become proficient at working through tasks. It is very important that study skills in general are taught to students from the time that we expect them to go home and study. We know this is very context dependent, so if it applies in your grade level, keep that in mind. Table 10.2 offers an example of a checklist for students to guide their work with integers.

Table 10.2 • Checklist for Solving Integer Problems

Record the problem.	Problem: −2 + 5
What is a real-life context for this problem? Tell a story about the problem.	Story: If I owed my brother $2 and I earned $5, I could pay my brother the $2, and I would have $3 left.
How will you model this problem? (number line, picture, red and white counters, integer beaded number line)	Model:
What is another way to model this problem? (number line, picture, red and white counters, integer beaded number line)	Model:
What is your solution? Does it make sense?	Solution: −2 + 5 = 3

Source: Pencil icon by Istock.com/Dann Fachrul; Earth icon by Istock.com/Turac Nvuzova.

Although there are mixed feelings about what note-taking in math class should look like, as learners ourselves, we all know the value of writing things down. As Maria Montessori famously said, "What the hand does, the mind remembers." That said, all note-taking is not created equal. As noted by Peter Liljedahl in *Building Thinking Classrooms in Mathematics* (2020), there is little benefit to students when they copy *our* notes. Instead, they should make their own notes as they consolidate their ideas and capture the key points that will prompt their memory. Exhibit 10.4 offers a sample of a note-taking structure that students could use following a lesson on slope in middle school. You can modify this for any topic, but it gives you the idea of how you can keep notes open-ended yet still help students to keep their thinking organized and support retrieval.

Exhibit 10.4 • Slope Note Catcher

Definition	
Types of Slope	
(Fill in the boxes below with words, drawings, or models to help you remember each type.)	
Positive	Negative
Zero	Undefined

Connections to SEL

Checklists and Rubrics

Checklists and rubrics offer a simple way to help students monitor their own learning. They promote ownership and accountability while also providing focus. The use of checklists and rubrics foster self-awareness because students reflect on their own learning performance and also how they feel about their success. They also foster self-management because students can use them to set goals and monitor their progress throughout tasks. They are a form of instant accountability so the student can self-evaluate and also show it to a peer and the teacher and get more feedback. Checklists and rubrics give everybody clarity so they can learn with a vision.

This can help students to feel calm, in control, and successful.

>
> ### Equity Check
> Different Scaffolds
>
> Remember that not all students can access paper organizers and templates, so including other modalities is necessary to ensure equity. Scaffolds that help organize information and resources are only effective when students can access them and use them. Some students prefer note scaffolds with many prompts and that guide them through step by step. Other students need basic outlines with a few prompts. Some will need this presented visually, while others may need it read audibly (either live or through an on-demand recording). The bottom line is that one size does not fit all, and being open to a variety of options will help ensure the success of all students.

Enhance Capacity for Monitoring Progress

Progress monitoring is an essential part of your practice, but what about students? If they are setting goals, they should be monitoring their progress to assess if they are headed in the right direction. Progress monitoring gives students feedback to determine the next steps. Research states that feedback should be explicit, timely, informative, clear, accessible, instructive, and student-friendly. CAST (2024) notes that offering plenty of "formative" feedback throughout a unit of study supports students because "learners can monitor their own progress effectively and to use that information to guide their own effort and practice."

Progress monitoring can take many forms. For example, the use of rubrics and checklists help students to pivot when they need to. The flexibility of Math Workshop and the versatility of universal design for learning come together in Math Workshop Plus to "provide opportunities to re-engage with content and revise work." This may involve finding a different modality, taking a step back, or realizing that they need to invest more effort. The rubric supports coaching conversations between you and your students, where the student does the work and evaluates themselves on the rubric before planning the next steps. Students then go back and continue working to make more progress.

Let's say, for example, a student turns in a word problem but all of the work is not included. You can redirect the student and talk about what the next moves can be. You might say something like this:

- "I like the way that you used the ten frames to model your thinking."
- "I like the way that you wrote the equation here at the bottom."
- "Can you look at the checklist and see if there is any part that you need to think more about?"

If the student finds something that they missed, they can go back and correct that. If the student does not find anything, you can engage in more questioning. Instead of saying, "You forgot to include _____," take the opportunity to model your own metacognitive process of using the rubric. This will help students learn how to use the rubric more effectively in the future. Also, continue to probe. Perhaps the student didn't complete a portion because they didn't know how, or they weren't sure if you would accept a picture instead of words. Maybe they couldn't read or comprehend what was being asked. To set students up for successful progress monitoring, we must make sure the feedback we offer is on point and advances the student. While often this feedback will be related to the content, remember that it can also be related to barriers in action and expression.

Conferring With Students

Conferring is the most common way we see teachers supporting students in monitoring their own progress. While it may sound unrealistic, especially if you are a middle school with multiple classes of different students, it's actually quick and effective. Think about the data you collect from benchmark assessments or exit tickets or student reflections. Any of these may be used to set goals and support progress monitoring. As part of the process, you plan these quick check-ins with students to see where they are and help them make adjustments to get where they're going. This could look like setting a new goal (after a happy dance) or sharing feedback about your observations related to time on task, organization, or the choice of workstations. During these conferences, you will not only give feedback but ask questions that get students to reflect on their progress. See Exhibits 10.5 and 10.6 for guides we use to capture these conversations. If you are someone who finds that you make things more difficult than they need to be, don't do it! These exchanges last for about 1 minute and are designed to build accountability, give succinct and clear feedback, and keep students moving along.

Exhibit 10.5 • **Math Conference Notes**

Math Conference Notes		
Date:		
Student:		
Type of Conference:		
Observations:	Compliments:	Teaching Points:
Goals:	Other Notes:	Next Meeting Date:

Source: Newton, 2024.

 Download this Math Conference Notes worksheet at https://companion.corwin.com/courses/MathWorkshopPlus.

Exhibit 10.6 • **Student Conferring Notes**

Date:	I am doing well with . . .	I am working on . . .

Source: Newton, 2024.

Student Choice in Assessment

Just as in other areas of Math Workshop Plus, students should have a choice regarding how to monitor their progress. In addition to having some flexibility in how they demonstrate proficiency in the assessments you administer, students should have the opportunity to choose different ways of providing evidence that they are progressing toward their goals (see Exhibit 10.6). It is important to talk with students about the importance of collecting evidence of learning along the way, so you know if they need more support before the end of the unit. Depending on what the assessment entails, you might remind students that there are many ways to show what they know. If you are up for experimenting with this, you may find that students are suddenly eager to be "assessed." In the example in Exhibit 10.7, consider how the product would or would not showcase what the student knows and how you might use something like this with your own spin.

Exhibit 10.7 • Assessment Tic Tac Toe

Assessment Tic Tac Toe Choose 3 ways to show that you know what we are learning!		
Blog post	Write a booklet	Social media post
Make a teaching poster	Free choice	Make a gameboard
Make a 5-question quiz with the answer key	Make a card or dice game	Make a podcast

Math Thinking Notebooks

What are Math Thinking Notebooks, and what role can they play in progress monitoring? Dr. Nicki has always talked about them as "Thinking Notebooks," putting the emphasis on the "Thinking." She tells students she only wants to see "evidence of thinking in their notebooks." Can you see how this could go sideways if you don't explicitly teach, model, and check what students are writing? We are laughing at what you may be envisioning at the moment, which likely involves some specific student who doesn't show any work (ever!) or one whose work appears, at first glance, to be one giant scribble!

If you implement Math Thinking Notebooks, you must actively plan for them. It is important to set aside time at the beginning of the year to organize them. This will look different depending on the grade you teach, but regardless of grade, we highly recommend making a sample and keeping it on display. In the lower grades, keep it simple. In the upper grades, you might tab off sections so that it is easy to distinguish between the parts. Again, the number of sections depends on the grade and the various components you intend to include.

Notebooks are a great tool to support ongoing assessment conversations. You can leave notes back and forth with the students about their work. These comments, as well as the content in the pages, can be used as springboards in math conferences. The notebooks are also great to demonstrate progress with word problems over time, capture weekly reflections, and generate conjectures during routines and energizers.

There are many ways to organize a notebook; these are just some:

- Insert a flap for students to keep track of where they are.
- Add a table of contents.
- Tab the notebook for efficient access across each section.
- Color-code the sections.

Connections to SEL

Reflection

Students should be encouraged to reflect on their journey. How they are feeling about themselves as mathematicians along the way. What is easy? What is challenging? What do they do when they are stuck? In what ways do they encourage themselves to persevere along the way? When and how do they seek out help? There should be space and opportunities for students to write about these things in their notebook.

Another great way to use notebooks is to have students comment and reflect on each other's work. They can do this using sticky notes. This should be scaffolded with language feedback prompts to ensure that students know how to read and write about the work of others in thoughtful, respectful, and helpful ways. This builds the art of perspective-taking and giving feedback respectfully.

> **Equity Check**
>
> Progress Monitoring
>
> Progress monitoring with an equity lens means leaning into different ways that students can show what they know. For example, some students struggle to explain in writing, but they can tell you the information. Alvarez (n.d., Slide 10) notes that "progress monitoring with equity in mind is achieved through varied formal assessments, group discussions, project-based learning and other instructional opportunities." Students are given tiered questions, tiered supports and scaffolds, vocabulary guides, and rubrics (Alvarez, n.d.).

Some students, especially those who aren't fluent with language, know the information being asked, but due to the language barrier, they are unable to show what they know. Whether through drawings, translation, or some other means, find a way to get academic information so you can accurately assess math. An important aspect of progress monitoring for multilingual learners is the types of questions they are being asked to answer. Try to pose questions that match their level of language so that they can answer. As you assess and monitor progress, continually check in with yourself to see if all students are being given an equitable opportunity to show what they know. If not, use this book as a guide to close that opportunity gap.

Challenge Exclusionary Practices

This is a tricky one, especially if you are someone who isn't comfortable rocking the boat. While we certainly get that, we ask you to take a moment to consider how it feels to be on the wrong side of someone's bias, discrimination, or exclusion, even if it is unintentional. Surely you can recall a time when you felt left out. Perhaps a group of friends from work went out after school and didn't invite you. If they told you afterward that they thought you had a meeting, would that feel different than if they said they just forgot? Maybe—or maybe not—because the fact remains that you missed out on a good time.

If you have made it to the last chapter of a book that is grounded in inclusive practices, we are guessing that you are committed to creating an environment where not only everybody feels as if they belong but also where nobody feels excluded. For this to become a reality, you need to set expectations, teach students how to communicate, and create a sense of mutual respect and accountability. Through community-building routines, the class can stay dialed in to what is happening. They have space to discuss things that have happened, how they felt, and how they think it should be resolved. They have a voice they feel comfortable using to hold each other accountable and stand up when something is not just.

Your students are a valuable, boots-on-the-ground resource to help you see the implications of words and actions, as well as the implications of *inaction*. At first, this may be tough to receive, but take it as feedback and use it spark your efforts to build the classroom community you aspire to. It will take time, skill, and possibly some professional learning to facilitate some of these conversations. You may have to leave your personal comfort at the door in the name of the greater good. Sometimes these are the difficult conversations, where classroom communities must name, explore, and address incidents of bias, discrimination, and exclusion happening in the community that are significant. Building an inclusive community is hard work but is well worth the effort.

Connections to SEL

Relationship Skills

Building inclusive communities requires that students build their relationship skills. It demands that students are honest, accepting, forgiving, and truthful about what is happening. Students become socially aware of what is happening because, through the meetings, your class will likely discuss some really tough issues. Students must grapple with their feelings, attitudes, and behaviors around what is happening in the classroom. As you continue on your own journey, you will hone your skills and increase your level of confidence in leading students on and ensuring that students are emotionally, physically, and academically safe.

Equity Check

Fostering Inclusive Communities

Challenging exclusionary practices is the epitome of taking visible steps to ensure equity and access in your classroom. This means that building an inclusive community is something that students must know you are serious about. It means that each year you will teach your students how to work, play, and live in a diverse world that is welcoming to everyone. It also means that you teach students how to spot bias and exclusionary practices when they happen and also how to address them in ways that resolve the issues. It's hard work, but it's worth it to ensure that in your classroom all students can learn without "fear of threat, humiliation, danger or disregard" (IDRA, 2020).

Say Bye-Bye to Barriers When You . . .

- Create a goal-setting environment
- Explicitly teach planning and organization skills
- Give opportunities to build perseverance
- Provide opportunities to pause and reflect
- Scaffold tasks and use checklists
- Give time to self-monitor and self-assess
- Progress monitor continuously
- Challenge exclusionary practices

Action Plan

Assess your current success with Design Options for Strategy Development. Download the Chapter 10 Self-Assessment at https://companion.corwin.com/courses/MathWorkshopPlus.

Try It! Increasing Access

Identify how you might start or tweak student access to tools, graphic organizers, diagrams, and manipulatives. Try it.

Connections

Share an artifact of something that you have done to address executive functions on social media. #mathworkshopPLUS.

Chapter 11

You've Got This!

> "There are only two ways to influence human behavior: You can manipulate it, or you can inspire it."
> — Simon Sinek
> (Columbia professor, author, and leadership guru)

Revisit Your Why

As we come to the conclusion of this book, we invite you to revisit what attracted you to it and why you chose to read it. We know that you have many demands on your time, tons on your plate, and that every minute is precious. So why *did* you read this book?

If you were looking for ideas to take your workshop to the next level, we feel certain that you got them! If you were looking for insight into the intersections of SEL and Math Workshop, or if you wanted a more concrete understanding of what UDL and equitable practices can look like in a math class, we feel confident that you got that too. While we intended to deliver those things, what we wanted most was to give you the gift of inspiration. You are the change agent, the difference maker, the heart and soul of the learning environment that students experience, and sometimes you forget how much you matter! If you invested your precious time to read, learn, reflect, and plan, we would feel like we let you down if we have not inspired you.

According to Thrash and Elliot (2003), people who are inspired share the following characteristics:

- Inspired people are more intrinsically motivated, which has a positive impact on their performance.
- Inspired people believe in their abilities and are optimistic.
- Inspired people use their inspiration as a springboard for creativity.

We know from experience that inspired people take action. Follow-up studies from other researchers noted that people who felt inspired showed an increase in progress toward their goals, which resulted in setting more inspired goals. That is our hope for you. We hope that you have been following through with the Action Items at the end of each chapter, experimenting with ideas, and seeing what works for you. If you haven't, you may feel overwhelmed; if that's the case, pause, breathe, and remember your why.

Go Slow to Go Fast

To paraphrase Simon Sinek, everyone starts with *why*, but only the great ones keep their why year after year. Keeping your "why" front and center will give you the motivation to turn your inspiration into actions that make a difference in the lives of your students. If you try to do too much at once, however, you may feel stressed, and that's not the point. Even if there are ideas in this book that excite you because you know they will make a difference for students, nobody said you have to do everything tomorrow! In fact, we encourage you to take baby steps. We have found that the secret to success is going slow now, so you can go fast later.

Try It!

Proactively Designing Options for Engagement, Representation, and Action & Expression

Without looking back at the chapters, use the following space to jot down five things you have already tried or are eager to try.

1. _____
2. _____
3. _____
4. _____
5. _____

Now take a moment to think about some specific students in your class who may be struggling. Ask yourself if the things on your list will help close opportunity gaps and create avenues for success for *those* students. If the answer is yes, great—you are ready to roll! If the answer is no, look back at the chapters and draft a different list. The point here is to try something that will yield immediate results for the students who need it most. This quick win will motivate you to keep going.

Much like the graphic organizers we suggested for your students, we want you to use your list as a guide to set goals, make a plan, and monitor your progress before setting new goals and starting the cycle again. This will not only create a road map but will give you permission to focus on a small number of specific changes, observe their impact, and assess their value. Hopefully, this will help you to slow down and not try to swallow the entire elephant at once.

You're a Learner Too

Maybe it comes with the territory of being a teacher, but we feel pretty sure that regardless of the reassurance we are offering, you have high expectations for yourself. So many of us have Type A personalities that drive us to strive for perfection, even when we know that is impractical, unrealistic, and, frankly, impossible! Nobody is perfect, and trying to be is not only exhausting but unnecessary. Please remember that when you are in learning mode, the only person who may be expecting you to be perfect is you.

Let us remind you that it's normal to feel uncomfortable when you're trying something new. And since growth only happens when you operate outside of your comfort zone, we want you to get comfortable with being uncomfortable! Be vulnerable. Make mistakes. Remind yourself that this is how your students feel every single day. If you're going to preach to them about the Power of Yet and growth mindset, you'd better walk your talk!

We wrote this book as an act of advocacy for the Rights of the Learner (Torres, 2020), and that applies to your students and also to you. The Rights of the Learner (RotL) state that your students, and *you,* have these rights:

- To be confused
- To claim a mistake and revise your thinking
- To speak, listen, and be heard
- To write, do, and represent only what makes sense to you

If you have ever been told to use a new program for math one year, literacy the next, science after that, or something similar, we know how frustrating it can be. It feels as if you can't get good at the first thing before the next thing comes along. We want to remind you that, while the ideas in this book are framed around math, they are content agnostic and may be applied in other areas. Take time to practice them where they fit and be patient with yourself as you learn.

Giving yourself grace as you learn through trial and error doesn't mean you're failing or not getting the hang of it. On the contrary, it's vital to see what works and what doesn't if you want to maintain your momentum and make permanent changes. The fact that you're trying is proof that you're invested, and your sustained effort will illustrate that you're in it to win it. This will show students that when people are inspired to grow and change, they back it up with effort, perseverance, and tenacity, even when it's hard. Let them see that you don't quit when something doesn't work and that every experience offers a chance to learn. Let them see that you believe they are worth it, and you won't give up.

We hope you know that you're worth it too. We applaud you for investing in yourself, investing in your students, and for doing everything in your power to create a learning environment that delivers on the promise of access and equity. Dr. Nicki and I see ourselves not as experts, but as learners. Our why is *you* and your students. Our why is the relentless pursuit of more accessible, equitable, and joyful mathematics for *all* students, *all* teachers, and anyone else involved in the learning process. *You* are what inspires us to keep going, to keep learning, to keep growing, and to keep creating. We hope that you experience the smiling faces, increased achievement, confidence, and pride that comes from doing this work. We cannot wait to see the photos and videos of your students and to hear about your experiences, so please remember to tag us when you post. We hope the collection of resources we have created will support your Math Workshop Plus journey and that you feel us cheering you on every step of the way!

References

Agarwal, P. K., Nunes, L. D., & Blunt, J. R. (2021). Retrieval practice consistently benefits student learning: A systematic review of applied research in schools and classrooms. *Educational Psychology Review, 33*(4), 1409–1453.

Alvarez, E. (n.d.). *Progress monitoring, feedback and engagement*. Office of Bilingual Education and Word Languages, NY State Education Department.

Assor, A., Kaplan, H., & Roth, G. (2002). Choice is good, but relevance is excellent: Autonomy-enhancing and suppressing teacher behaviors predicting students' engagement in schoolwork. *British Journal of Educational Psychology, 72*(2), 261–278.

Atkinson, R., Derry, S., Renkl, A., & Wortham, D. (2000, June). Learning from examples: Instructional principles from the worked examples research. *Review of Educational Research, 70*(2). https://doi.org/10.3102/00346543070002181

Banks, J. A., & Banks, C. M. (Eds). (2007). *Multicultural education: Issues and perspectives* (6th ed.). Wiley.

Barroso, C., Ganley, C. M., McGraw, A. L., Geer, E. A., Hart, S. A., & Daucourt, M. C. (2021). A meta-analysis of the relation between math anxiety and math achievement. *Psychological Bulletin, 147*(2), 134.

Bishop, P. A., & Pflaum, S. W. (2005). Middle school students' perceptions of social dimensions as influencers of academic engagement. *RMLE Online, 29*(2), 1–14.

Bishop, R. S. (1990). Mirrors, windows, and sliding glass doors. *Choosing and Using Books for the Classroom, 6*(3). Summer.

Bledsoe, T. S., & Baskin, J. J. (2014). Recognizing student fear: The elephant in the classroom. *College Teaching, 62*(1), 32–41.

Boaler, J. (2008). *What's math got to do with it?: How parents and teachers can help children learn to love their least favorite subject*. Penguin.

Boaler, J. (2014). *Fluency without fear.* https://www.youcubed.org/evidence/fluency-without-fear/

Boaler, J. (2015). *Mathematical mindsets: Unleashing students' potential through creative math, inspiring messages and innovative teaching*. Wiley.

Boaler, J. (2016). *How you can be good at math, and other surprising facts about learning* [Video]. TED Conferences. https://www.youtube.com/watch?v=3icoSeGqQtY

Boulet, G. (2007). How does language impact the learning of mathematics? Let me count the ways. *Journal of Teaching and Learning, 5*(1), 1–12.

Braselton, S., & Decker, B. C. (1994). Using graphic organizers to improve the reading of mathematics. *The Reading Teacher, 48*, 276–281.

Bronfenbrenner, U. (Ed.). (2005). *Making human beings human: Bioecological perspectives on human development.* Sage.

Brownell, J. (2023). *What is a number path and how does it build number relationships?* https://earlymath.erikson.edu/number-path/

Bruner, J. S., Goodnow, J. J., & Austin, G. A. (1956). *A study of thinking.* Chapman & Hall.

Bushart, B. (2023). *Numberless word problems.* https://numberlesswp.com/.

CASEL. (2020). *CASEL's SEL framework.* https://casel.org/casel-sel-framework-11-2020/

CAST. (2024). *The UDL guidelines.* https://udlguidelines.cast.org/

Center on the Developing Child, Harvard University. (n.d.). *What is executive function? And how does it relate to child development?* https://developingchild.harvard.edu/resources/what-is-executive-function-and-how-does-it-relate-to-child-development/#:~:text=The%20phrase%20%E2%80%9Cexecutive%20function%E2%80%9D%20refers,focused%20despite%20distractions%2C%20among%20others

Code, J. (2020, February). Agency for learning: Intention, motivation, self-efficacy and self-regulation. In *Frontiers in education* (Vol. 5, p. 19). Frontiers Media SA.

Cohen, E. G. (1994). *Designing groupwork: Strategies for heterogeneous classrooms* (Rev. ed.). Teachers College Press.

Cohen, E. G., & Lotan, R. A. (1995). Producing equal-status interaction in the heterogeneous classroom. *American Educational Research Journal, 32*(1), 99–120. https://doi.org/10.3102/00028312032001099

Dabrowski, J., & Marshall, T. R. (2018). *Motivation and engagement in student assignments: The role of choice and autonomy—Equity in motion.* https://eric.ed.gov/?id=ED593328

Darling-Hammond, L., & Cook-Harvey, C. M. (2018). *Educating the whole child: Improving school climate to support student success.* Learning Policy Institute. https://doi.org/10.54300/145.655

Dean, C. B., Stone, B. J., Hubbell, E., & Pitler, H. (2012). *Classroom instruction that works: Research-based strategies for increasing student achievement* (2nd ed.). ASCD.

Dewey, J. (1938). *Experience and education.* Macmillan.

Dixon, J. (2018). *Providing scaffolding just in case: Five ways we undermine efforts to increase student achievement (and what to do about it!)*. http://www.dnamath.com/blog-post/five-ways-we-undermine-efforts-to-increase-student-achievement-and-what-to-do-about-it-part-3-of-5/

Dixon, J., Brooks, L., & Carli, M. (2019). *Making sense of mathematics for teaching the small group*. Solution Tree Press.

Donovan, M. S., & Bransford, J. D. (2005). *How students learn—Science in the classroom*. National Academy Press.

Doubet, K. (2022). *The flexibly grouped classroom: How to organize learning for equity and growth*. ASCD.

Driscoll, M. J., Nikula, J., & DePiper, J. N. (2016). *Mathematical thinking and communication: Access for English learners*. Heinemann.

Dusenbury, L., Yoder, N., Dermody, C., & Weissberg, R. (2020). *An examination of K–12 SEL learning competencies/standards in 18 states*. Frameworks Briefs. CASEL.

Dweck, C. S. (2007). *Mindset: The new psychology of success*. Ballantine Books.

Emeny, W. G., Hartwig, M. K., & Rohrer, D. (2021). Spaced mathematics practice improves test scores and reduces overconfidence. *Applied Cognitive Psychology, 35*, 1082–1089.

Erikson, E. H. (1963). *Childhood and society* (2nd ed.). Norton.

Farrington, C. A. (2013). *Academic mindsets as a critical component of deeper learning*. Consortium on Chicago School Research.

Fisher, D., & Frey, N. (2014). *Checking for understanding: Formative assessment techniques for your classroom*. ASCD.

Garskof, J. (n.d.). *6 life skills kids need for the figure*. https://www.scholastic.com/parents/family-life/creativity-and-critical-thinking/learning-skills-for-kids/6-life-skills-kids-need-future.html

Gernsbacher, M. A., Soicher, R. N., & Becker-Blease, K. A. (2020). Four empirically based reasons not to administer time-limited tests. *Translational Issues in Psychological Science, 6*(2), 175–190. https://doi.org/10.1037/tps0000232

Goldberg, P. (2003). Using metacognitive skills to improve 3rd graders' math problem solving. *Focus on Learning Problems in Mathematics, 5*(9), 12–27.

Gómez, C. L. (2010). Teaching with cognates. *Teaching Children Mathematics, 16*(8), 470–474.

Gonzalez, N., Moll, L. C., & Amanti, C. (Eds.). (2005). *Funds of knowledge*. Routledge.

Hammond, Z. (2014). *Culturally responsive teaching and the brain: Promoting authentic engagement and rigor among culturally and linguistically diverse students*. Corwin.

Hammond, Z. (2018). Culturally responsive teaching puts rigor at the center. *Focus Equity, 39*(5). www.learningforward.org

Harris, P. (2023). *Problem strings*. https://www.mathisfigureoutable.com/blog/problem-string

Harvard University. (n.d.). *Center on the Developing Child: Building the brain's "air traffic control" system—How early experiences shape the development of executive function.* https://developingchild.harvard.edu/resources/working-paper/building-the-brains-air-traffic-control-system-how-early-experiences-shape-the-development-of-executive-function/

Hattie, J., Biggs, J., & Purdie, N. (1996). Effects of learning skills interventions on student learning: A meta-analysis. *Review of Educational Research, 66*(2), 99–136. https://doi.org/10.2307/1170605

Hess, K. (2023). *Rigor by design, not chance: Deeper thinking through actionable instruction and assessment*. ASCD.

Hough, L. (2022). A space for joy: Educators talk about the impact COVID has had on school happiness. https://www.gse.harvard.edu/ideas/ed-magazine/22/05/space-joy#:~:text=%E2%80%9CInstead%20of%20taking%20pleasure%20from,developing%20relationships%20with%20their%20teachers.%E2%80%9D

Hunter, M. (1969). *Motivation theory for teachers*. Corwin.

Immordino-Yang, M. H. (2016). *Emotions, learning, and the brain: Exploring the educational implications of affective neuroscience*. W. W. Norton.

Institute of Education Sciences (IES). (2021, March). *Assisting students struggling with mathematics: Intervention in the elementary grades*. U.S. Department of Education. https://ies.ed.gov/ncee/wwc/Docs/PracticeGuide/WWC2021006-Math-PG.pdf

Intercultural Development Research Association (IDRA). (2020). *Six goals of educational equity*. https://www.idra.org/equity-assistance-center/six-goals-of-education-equity/

Inzlicht, M., Shenhav, A., & Olivola, C. Y. (2018). The effort paradox: Effort is both costly and valued. *Trends in Cognitive Science, 22*(4), 337–349. doi:10.1016/j.tics.2018.01.007

ISTE. (n.d.). *ISTE standards: For students*. https://iste.org/standards/students

Johnson, D., & Johnson, R. (1998). *Learning together and alone: Cooperative, competitive and individualistic learning* (5th ed.). Pearson.

Kagan, S. (1994). *Cooperative learning*. Author.

Kannas, K., Colombo, J., & Wyss, N. (2010). Now, pay attention! The effects of instruction on children's attention. *Journal of Cognition and Development, 11*(4), 509–532. https://www.ncbi.nlm.nih.gov/pmc/articles/PMC3015160/

Kelemanik, G., Lucenta, A., & Creighton, S. J. (2016). *Routines for reasoning: Fostering the mathematical practices in all students*. Heinemann.

Kenney, J. (2022). *Executive functioning skills and math success*. https://www.lynnekenney.com/post/executive-function-skills-and-math-what-s-the-relationship

Kluth, P. (2020). *Universal design daily: 365 ways to teach, support, & challenge all learners*. Author.

Krashen, S. (1982). *Principles and practice in second language acquisition*. Pergamon Press.

Ladson-Billings, G. (1995). Toward a theory of culturally relevant pedagogy. *American Educational Research Journal, 32*(3), 465–491.

Lambert, R. (2021). The magic is in the margins: UDL math. *Mathematics Teacher: Learning and Teaching PK–12, 114*(9), 660–669.

Lambert, R., Imm, K., Schuck, R., Choi, S., & McNiff, A. (2021). "UDL is the what, design thinking is the how": Designing for differentiation in mathematics. *Mathematics Teacher Education and Development, 23*(3), 54–77.

Lange, J. (2021). *The importance of using manipulatives in math class*. Master's theses and Capstone projects, Northwestern College, IA. https://nwcommons.nwciowa.edu/cgi/viewcontent.cgi?article=1291&context=education_masters

Leong, Y. H., Ho, W. K., & Cheng, L. P. (2015). Concrete-pictorial-abstract: Surveying its origins and charting its future. *Mathematics Educator, 16*(1), 1–18.

Leppänen, U., Niemi, P., Aunola, K., & Nurmi, J.-E. (2006). Development of reading and spelling Finnish from preschool to Grade 1 and Grade 2. *Scientific Studies of Reading, 10*(1), 3–30. https://doi.org/10.1207/s1532799xssr1001_2

Letourneau, C., & Freed, G. (2000). *Guideline: Provide equivalent alternatives to auditory and visual content*. https://www.w3.org/WAI/wcag-curric/gid2-0.htm

Library City University of Seattle. (2024). *Accessibility practices for charts, graphs, tables, and infographics*. https://library.cityu.edu/researchguides/accessibility/chartsandtables

Liljedahl, P. (2020). *Building thinking classrooms in mathematics, grades K–12*. Corwin.

Lipkin, P. H., & Okamoto, J., Council on Children with Disabilities and Council on School Health, Norwood Jr., K. W., Adams, R. C., Brei, T. J., . . . & Young, T. (2015). The individuals with disabilities education act (IDEA) for children with special educational needs. *Pediatrics, 136*(6), e1650–e1662.

Mader, J. (2022). *Want resilient and well-adjusted kids? Let them play.* The Hechinger Report. https://hechingerreport.org/want-resilient-and-well-adjusted-kids-let-them-play/

Malamed, C. (n.d.). *Six ways to use examples and nonexamples to teach concepts*. The eLearning Coach. https://theelearningcoach.com/elearning_design/examples-and-nonexamples/

Marzano, R. J. (2001). *Classroom instruction that works: Research-based strategies for increasing student achievement*. ASCD.

Mayer, R. E. (2009). *Multimedia learning* (2nd ed.). Cambridge University Press.

McDonald, S. W., Kehler, H. L., & Tough, S. C. (2018). Risk factors for delayed social-emotional development and behavior problems at age two: Results from the All Our Babies/Families (AOB/F) cohort. *Health Science Reports, 1*(10), e82. https://doi.org/10.1002/hsr2.82

McFarland, J., Hussar, B., Zhang, J., Wang, X., Wang, K., Hein, S., . . . & Barmer, A. (2019). The condition of education 2019. NCES 2019-144. *National Center for Education Statistics*.

McGinn, K., Lange, K., & Booth, J. (2015). A worked example for creating worked examples. *Mathematics Teaching in the Middle School, 21*(1), 27–33.

Merrill, S., & Gonser, S. (2021). *The importance of student choice across all grade levels.* Edutopia. https://www.edutopia.org/article/importance-student-choice-across-all-grade-levels/.

Meyer, A., Rose, D. H., & Gordon, D. (2014). *Universal design for learning: Theory and practice.* CAST.

Meyer, D. (n.d.). *Sugar: A 3-act task.* https://threeacts.mrmeyer.com/sugarpackets/

Midwest Comprehensive Center. (2018). *Student goal setting: An evidence-based practice.* https://files.eric.ed.gov/fulltext/ED589978.pdf

Mihalko, J. C. (1978). The answers to the prophets of doom: Mathematics teacher education. In D. B. Aichele (Ed.), *Mathematics teacher education: Critical issues and trends* (pp. 36–41). National Education Association.

Miller, D. (2022). *The intersection of equity and planning.* https://www.foursquareitp.com/blog/the-intersection-of-equity-and-planning/

Missouri State Admin. (2017). Accessibility updates. https://blogs.missouristate.edu/accessibility/2017/12/13/accessibility-dont-use-color-alone-to-convey-content/#:~:text=The%20accessible%20solutions,%2C%20identifiers%2C%20symbols%20or%20patterns

Montessori, M. (1967). *The discovery of the child.* Ballantine Books.

Moser, J. S., Schroder, H. S., Heeter, C., Moran, T. P., & Lee, Y. H. (2011). Mind your errors: Evidence for a neural mechanism linking growth mind-set to adaptive posterror adjustments. *Psychological Science, 22*(12), 1484–1489.

Mostafa, K. (2017). A kind of cognitive bridge: Using graphic organizers to support learning. *International Journal of Educational Development, 56,* 1–8. https://doi.org/10.xxxx/educational.dev.2017

Murawski, W. W., & Novak, K. (Eds.). (2019). *What really works with Universal Design for Learning.* Corwin.

Najarro, I. (2023). *What is translanguaging and how can it support dual-language learners?* EdWeek. https://www.edweek.org/teaching-learning/what-is-translanguaging-and-how-is-it-used-in-the-classroom/2023/07

National Council of Teachers of Mathematics (NCTM). (2014). *Principles to actions: Ensuring mathematical success for all.* NCTM.

National Research Council. (2001). *Adding it up: Helping children learn mathematics.* J. Kilpatrick, J. Swafford, & B. Findell (Eds.). National Academy Press.

Newton, N. (2016). *Math running records in action.* Routledge.

Newton, N. (2022). *Math workshop booklet.* Newton Education Solutions.

Newton, N. (2023). *Accelerating K–8 math instruction.* Teachers College Press.

Newton, N. (2024). *Math fact fluency playground.* www.mathfactfluencyplayground.com

Newton, N., Mello, A., & Record, A. (2019) *Fluency doesn't just happen with addition and subtraction.* Routledge.

Newton, N., Mello, A., & Record, A. (2024). *Fluency doesn't just happen with multiplication and division.* Routledge.

Nguyen, N. (2021). *Feedback and universal design for learning (Feedback series).* https://feedbackfruits.com/blog/feedback-and-universal-design-for-learning

Paley, V. (1993). *You can't say, you can't play.* Harvard University Press.

Parrish, S. (2010). *Number talks: Helping children build mental math and computation strategies, grades K–5.* Math Solutions.

Piaget, J. (1959). *The language and thought of the child* (3rd ed.) (M. Gabain & R. Gabain, Trans.). Routledge.

Pierson, R. (2013, May 3). *Every kid needs a champion* [Video]. TED Conferences. https://www.bing.com/videos/riverview/relatedvideo?q=Rita+Pierson+2013+TED+Talk&&mid=937FFD13E5AB23319662937FFD13E5AB23319662&FORM=VAMGZC

Prensky, M. (2001). Digital natives, digital immigrants part 2: Do they really think differently? *On the Horizon, 9*(6), 1–6.

Prepared Parents. (n.d.). *These skills prepare kids for any future.* https://preparedparents.org/editorial/universal-skills-for-kids-to-succeed/

Read, M. (2019). Designing with color in the early childhood education classroom: A theoretical perspective. *Creative Education, 10*, 1070–1079. https://doi.org/10.4236/ce.2019.106080

Resnik, M. (1981). Mathematics as a science of patterns: Ontology and reference. *Noûs, 15*(4), 529–550. doi:10.2307/2214851

Rubenstein, R. N., & Thompson, D. R. (2002). Understanding and supporting children's mathematical vocabulary development. *Teaching Children Mathematics, 9*(2), 107–112.

Sammons, L. (2009). *Guided math: A framework for mathematics instruction—A framework for mathematics instruction.* Teacher Created Materials.

Schleppegrell, M. J. (2004). 10 technical writing in a second language: The role of grammatical metaphor. *Analysing Academic Writing*, 172.

Schmitt, B. (2024). Attention deficit hyperactivity disorder (ADHD). *Pediatric Patient Education*. Schmitt Pediatric Guidelines LLC.

Schoenfeld, A. H. (2020). Mathematical practices, in theory and practice. *ZDM, 52*(6), 1163–1175.

Sinek, S. (2011). *Start with why: How great leaders inspire everyone to take action.* Portfolio/Penguin.

Smith, C. (2021). *13 effective study strategies to help students learn.* https://www.kqed.org/mindshift/57644/13-effective-study-strategies-to-help-students-learn

Smith, M. S., Hillen, A. F., & Catania, C. L. (2007). Using pattern tasks to develop mathematical understandings and set classroom norms. *Mathematics Teaching in the Middle School, 13*(1), 38–44.

Stern, J., Ferraro, K., Duncan, K., & Aleo, T. (2021). *Learning that transfers: Designing curriculum for a changing world.* Corwin.

Su, F. E. (2017). Mathematics for human flourishing. *American Mathematical Monthly, 124*(6), 483–493.

Sun, K. L., Baldinger, E. E., & Humphreys, C. (2018). Number talks: Gateway to sense making. *Mathematics Teacher, 112*(1), 48–54.

Thorpe, C. (2010). *Promoting academic achievement in the middle school classroom: Integrating effective study skills instruction.* https://files.eric.ed.gov/fulltext/ED510601.pdf

Thrash, T. M., & Elliot, A. J. (2003). Inspiration as a psychological construct. *Journal of Personality and Social Psychology, 84*(4), 871.

Tomlinson, C. (1999). *The differentiated classroom responding to the needs of all learners.* ASCD.

Torres, O. G. (2020, August 12). *Equity in education webinar series: Rehumanizing schools—Rights of the learner.* [Video]. YouTube. https://www.youtube.com/watch?v=_UndpNUCAqw

United Nations. (1989). *Convention on the rights of the child.* https://www.ohchr.org/en/instruments-mechanisms/instruments/convention-rights-child

Usanmaz, O. (2024). *Understanding diversity, equity, inclusion, and belonging (DEIB) growing importance in 2024.* https://www.qooper.io/blog/what-is-diversity-equity-inclusion-and-belonging#:~:text=Diversity%20definition%20simply%20means%20the,and%20part%20of%20a%20community.

Vilenius-Tuohimaa, P., Aunola, K., & Nurmi, J. (2008) The association between mathematical word problems and reading comprehension. *Educational Psychology, 28*(4), 409–426.

Vygotsky, L. S. (1962). *Thought and language.* MIT Press.

Vygotsky, L. S. (1978). *Mind in society: The development of higher psychological processes.* Harvard University Press.

Wakefield, D. V. (2000, September). Math as a second language. In *The educational forum* (Vol. 64, No. 3, pp. 272–279). Taylor & Francis.

Wakefield, E. M., Congdon, E. L., Novack, M. A., Goldin-Meadow, S., & James, K. H. (2019). Learning math by hand: The neural effects of gesture-based instruction in 8-year-old children. *Attention, Perception, & Psychophysics, 81,* 2343–2353.

Walck-Shannon, E., Rowell, S., & Frey, R. (2021). *To what extent do study habits relate to performance?* Knight, J. (Monitoring Editor). https://www.lifescied.org/doi/10.1187/cbe.20-05-0091

Ward, C. (2020). What are normal attention spans for children? Retrieved on March 10, 2024, from https://www.kids-houston.com/normal-attention-spans-for-kids/

Welsh, R. O., & Little, S. (2018). The school discipline dilemma: A comprehensive review of disparities and alternative approaches. *Review of Educational Research, 88*(5), 752–794.

Wiggins, G. (2012). Seven keys to effective feedback. Retrieved on Jan. 2, 2024, from https://www.ascd.org/el/articles/seven-keys-to-effective-feedback

Williams, S. (2023). How to give feedback: What's equity got to do with it? Retrieved on Jan. 2, 2024, from https://www.snhu.edu/about-us/newsroom/career/how-to-give-feedback

Willis, J. (2007). The neuroscience of joyful education. Retrieved Sept. 24, 2024, from https://ascd.org/el/articles/the-neuroscience-of-joyful-education

Index

Absolute value, 185, 186 (figure)
Academic language, 140
Americans with Disabilities Act (ADA), 10
Artificial intelligence (AI), 50, 233
Assistive technology, 202, 217, 222

Boaler, J., 90
Braselton, S., 158
Building knowledge
   advanced organizers, 174–177, 175–177 (figure)
   cues and prompts, 187–188, 188 (figure)
   examples/nonexamples, 185–186, 185–186 (table)
   graphic organizers, 179–181, 179–181 (figure)
   information processing skills, 169
   interactive graphic organizers, 181, 182 (figure)
   interactive models, 191–193, 192–193 (figure)
   learners, 189
   Math Workshop Plus, 178
   maximize transfer and generalization, 195–198, 197 (table), 198 (figure)
   organizational methods and approaches, 189–191, 190 (figure), 191 (figure)
   prior knowledge to new learning, 170–172, 171 (table)
   science of patterns, 178
   support information processing strategies, 193–194
   triangle vocabulary, 182–183, 183 (table)
   unit organizers, 183, 184 (figure)
   visual imagery, 172–174, 173 (figure), 174 (figure)

Center for Applied Special Technology (CAST), 11, 13, 16, 45, 49, 50, 54, 58, 81, 82, 100, 111, 141, 170, 179, 180, 183, 189, 202, 253, 258, 260, 265
Classroom learning environment (CLE), 143
Classrooms, 1
   assessment, 217
   auditory processing issues, 11
   community, 28, 79, 80, 271
   culture and climate, 18
   early childhood, 120
   elements, 3
   first-grade, 239
   learning environment, 14
   Math Workshop Plus, 78, 103
   orientation, 146
   social and physical environment, 54
   third-grade, 51
   tool kit, 143
Cohen, E. G., 78
Coherence principle, 113
Collaborative for Academic Social and Emotional Learning (CASEL), 18, 20 (figure), 21, 25, 46, 100
Common Core State Standards (CCSS), 140
Community-building routines, 270

Concrete-representational-abstract (CRA) model, 229
Cook-Harvey, C. M., 19, 33
COVID-19 pandemic, 244
Cultural and social relevance, 50

Dabrowski, J., 213
Darling-Hammond, L., 19, 33
Decker, B. C., 158
Digital Rekenreks, 191, 192 (figure)
Digital tools, 57, 165, 167, 239, 240, 242, 243
Diversity, 1–2, 27, 52, 79, 133, 134, 160, 257
Diversity, equity, inclusion, and belonging (DEIB), 81
Dixon, J., 247
Doubet, K., 217
Driscoll, M. J., 140
Dweck, C. S., 88

Effective supports, 245
Efficient, effective, and easy (3 Es), 204
Effort meter, 255–256 (figure), 255–257
Elliot, A. J., 274
Embodiment principle, 120, 121 (figure)
Emotional capacity
   adulthood, 91
   beliefs and motivation, 92, 92–94 (table)
   cultivate empathy, 109
   evidence-based, 90
   fear, anxiety and stress, 87
   individual and collective reflection, 103–105, 105 (table)
   math disposition, 88
   math education community, 89
   mistakes poster, 89, 89 (figure)
   negative experiences, 87
   normalize mistakes, 106, 106–108 (table)
   principal, 90
   responses and interactions, 99–102
   restorative practices, 109
   self and others, develop awareness of, 95–99, 96–98 (figure)
   student learning, 91

"Enactive-iconic-symbolic" model, 229
Environmental cues, 187
Equivalent fractions
   fraction bars modeling, 241–242, 242 (figure)
   pattern blocks modeling, 241–242, 242 (figure)
Esti-Mystery, 206–209, 207 (figure), 208 (figure)
Evidence-based strategy, 160
Executive-function skills, 251
Expression and communication
   build fluencies with graduated support, 244–249, 245–247 (table)
   C-R-A progression, 239
   digital media, 234
   digital tools, 239–241
   equivalent fractions, 241–242, 242 (figure)
   general educators, 239
   learning, 233–234
   materials and physical environment, 228
   modes, 249–250
   multiple media, 229–233, 232 (figure)
   number line, 241 (figure)
   paper-folding practice. *See* Paper-folding practice
   physical manipulatives, 239
   special educators, 239
   supporting students, 227
   tools and practices, 228
   variations, 227
   virtual manipulatives, 234, 239

Fisher, D., 187
Frayer model, 149, 182
   absolute value, 185, 186 (figure)
   polygons, 185, 186 (figure)
Frey, N., 187

Gestural cues, 187, 188
Goal planner, 254
Goal setting
   confidence and ownership, 63, 63 (table)
   meaning and purpose, 62–63

monitoring progress, 67, 67–68 (table)
multiplication chart, 64, 64 (figure), 65 (figure)
multiplication pathways, 66, 66 (figure)
plan development, 66–67
recognizing and celebrating goals, 68–69
starting point identification, 66
Google Classroom, 41, 225–226
Graphic organizers, 125, 129, 130, 165, 174–177, 179–181, 179–181 (figure), 183 (table), 187, 189, 196, 198, 258, 261, 262, 275
Guided Math: A Framework for Mathematics Instruction (Sammons), 4
Guided Math groups, 217–219
    gestural cue, 188
    Reference Folder, 181, 248
    small-group games, 38
    whole-group opportunities, 194
    and workstations, 8, 194

Hammond, Z., 195
Hexagons, 185, 185 (figure)

Image principle, 120
Inclusive community, 271
Individual education plans (IEPs), 222
Individuals with Disabilities Education Act (IDEA), 222
Ineffective supports, 245
Interaction
    access to technology, 225–226
    access to tools, 222–225, 223–224 (figure)
    artifacts and accountability, 214, 215 (table)
    assistive technology, 222
    the debrief, 219–221, 220–221 (table)
    Guided Math groups, 202, 217–219
    learning environment, 201
    mini lesson, 209–210 (table), 209–212, 211–212 (figure)

Number Talks, 203–205 (figure), 203–206, 205–206 (table)
    other launch routines, 206–209, 207 (figure), 208 (figure)
    purposeful practice activities, 215–216
    routines and energizers, 202–203
    student communication, 216–217
    students learn and process information, 201
    traditional math lesson, 201
    workstations, 202, 212–214, 213 (table)
Interactive language, 151
Interests & Identities, welcoming
    address biases, threats, and distractions, 54–56
    Authentic Connections Support Equity, 52
    choice and autonomy, 39–41
    grade-level standard, 34
    increasing authenticity, 51–52
    increasing relevance, 49–50
    increasing value, 50–51
    integrating routines, 46–49
    learning environments, 33
    noise levels, 56
    nurture joy and play, 52–54
    problem-solving approaches, 35, 36 (table)
    student-made board games, 42–45
    students experience, 34
    timed tests, 57–58
    workstation playlist and option, 37–38

Johnson, D., 78
Johnson, R., 78
Just-in-time scaffolds, 247

Kagan, S., 78
Kindergarten, 36 (table), 50, 53, 55, 70 (figure), 71 (figure), 80, 159, 172

Lambert, R., 11
"Language heavy" problems, 139, 141
Language & symbols

address biases, 165
anchor charts, 156–157, 156–157 (table)
classroom as a tool kit, 143
cognates, 160–161, 161 (table)
comprehensible input, 161
frames/sentence stems, 143–145, 144–145 (figure)
gestures, 161, 162–163 (figure)
highlighting roots and prefixes, 154–155, 154–155 (figure)
implications, 139
individual reference folders, 150–153, 152–153 (table)
math dictionaries, 148
math textbook/state assessment, 139
multiple media, 165–167
multiplication, 164
procedural fluency, 139
repetition, 163, 164 (table)
text and mathematical notation, 158–160, 159–160 (table)
visual vocabulary cards, 145–146, 145 (figure), 146 (figure)
vocabulary workstation, 141–143, 149–150
word wall, 146–148, 147 (figure)
Learning environment, 2, 5, 10, 14, 18, 25–27, 31, 33, 53, 54, 79, 95, 109, 111, 121, 131, 141, 142, 201, 216, 229, 252, 257, 273–274, 276
Lotan, R. A., 78

Marshall, T. R., 213
Marzano, R. J., 34
Mathematical Mindsets (Boaler), 89
Math Running Records (Newton), 57
Math Workshop
　debrief, 8–10, 9 (table)
　elements, 3
　Guided Math groups, 8
　instructional model design, 7
　Math Workshop Plus, 4–5
　mini lesson, 7–8

one-size-fits-all approach, 31
opener, 7
practices and challenges, 5
social and emotional competencies, 5, 29
social and emotional learning (SEL). *See* Social and emotional learning (SEL)
student-centered instructional approach, 3
Universal Design for Learning (UDL). *See* Universal Design for Learning (UDL)
workstations, 8
Mayer, R. E., 112. *See also Multimedia Learning* (Mayer), principles
Microsoft 365, 225–226
Mindset: The New Psychology of Success (Dweck), 88
Modality principle, 119
Moser, J. S., 89
Multimedia Learning (Mayer), principles
　accessibility notes, 122, 123 (table)
　coherence principle, 113
　embodiment principle, 120, 121 (figure)
　image principle, 120
　modality principle, 119
　multimedia principle, 119
　personalization principle, 120
　pre-training principle, 117–119, 119 (table)
　redundancy principle, 115
　segmenting principle, 115–117, 116–117 (figure)
　signaling principle, 113–114, 114 (figure)
　spatial contiguity principle, 115
　temporal contiguity principle, 115
　voice principle, 120
Multilingual learners (ML), 141, 160, 270
Multimedia principle, 119

National Council of Teachers of Mathematics (NCTM), 26
Neurodiversity, 1, 99, 134
Number strings, 47–49

Paper-folding practice
   color-coded mat, 236, 236 (figure)
   mat with model fold lines, 236, 237 (figure)
   symbol mat with emojis, 236, 237 (figure)
   symbol mat with shapes, 236, 236 (figure)
Partner Reflection Checklist, 126, 127 (figure)
Partner Talk Rubric, 128, 128 (figure)
Perception
   accessibility notes, 122, 123 (table)
   captioning slides and videos, 129–130
   classroom, 125, 126 (figure)
   diversity, 133–136, 134–136 (figure)
   learning environment, 111
   paper tables and charts, 124
   verbal instructions, 126–128, 127 (figure), 128 (figure)
   visual information, alternatives for, 130–131 (figure), 130–132
Personalization principle, 120
Physical diversity, 134
Polygons, 185, 186 (figure)
Problem-solving
   approaches, 35, 36 (table)
   checklist, 260
   templates, 258–259
Progress monitoring
   assessment, student choice, 268
   challenge exclusionary practices, 270–271
   conferring with students, 266–268
   "formative" feedback, 265
   Math Thinking Notebooks, 268–270
   metacognitive process, 266
   modality, 265
   students feedback, 265

Question Organizer, 172

Redundancy principle, 115
Reframing differences, 2
Relationship building, 45
Relationship skills, 20, 21, 45, 53, 68, 80, 100, 206, 211, 218, 271
Research-based tool, 189
Responsible decision-making, 20–22, 57, 217
Riddles and Number Talks, 46
Rights of the Learner (RotL) state, 276
Rubenstein, R. N., 141

"Salad bar" model, 223
Schleppegrell, M. J., 140
Science of patterns, 178
Seesaw, 225–226
Segmenting principle, 115–117, 116–117 (figure)
Self-awareness, 20, 21, 57, 68, 83, 90–91, 100, 104, 121, 178, 195, 198, 206, 217, 218, 257, 264
Self-efficacy, 25, 45, 83, 244, 248, 249
Self-management, 20, 21, 42, 83, 90–91, 100, 104, 188, 195, 206, 217, 218, 230, 244, 249, 257, 264
Self-motivation, 45, 83, 100
Self reflection, 45, 57, 74, 188, 257
Sixth-grade, 36 (table)
Social and emotional learning (SEL), 4
   access information, 132
   checklists and rubrics, 264
   classrooms and learning experiences, 18
   competencies, 18
   debrief, 221
   educational equity and excellence, 18
   education and human development, 18
   equity, 26–28
   fostering collaboration, 78
   fostering relationship skills, 80
   joy and play foster harmonious relationships, 53
   language supports, 141
   lean into others' perspectives, 52
   learning environment, 26
   math disposition, 25
   mini lessons, 210

normalizing tools, 225
ownership of learning, 68
peer work, 261
personal SEL goals, 102
prior knowledge, 178
receiving feedback, 83
reflection, 269
relationship building, 45
relationship skills, 21, 271
responsible decision-making, 21–22
routines, 206
scaffolding, 248
self-awareness, 21, 74, 90, 121, 257
self-efficacy, 25, 45
self-management, 21, 42, 90, 195, 257
self-motivation, 45
self reflection, 45, 67
skills, 22, 25–26
social awareness, 21
teaching study skills, 198
Social awareness, 20, 21, 52, 53, 78, 80, 206, 211
Spatial contiguity principle, 115
Standards of Mathematical Practice (SMP), 140
Strategy development
  anticipate and plan, 257–258
  estimate effort, resources and challenges, 254–257, 255–256 (figure)
  executive-function skills, 251
  goal setting, 253
  learning environment, 251–252
  organize information and resources, 262–265, 263 (table)
  problem-solving checklist, 260
  problem-solving templates, 258–259
  progress monitoring. *See* Progress monitoring
  templates and visual organizers, 261–262
Student-centered instructional approach, 3
Student-made board games, 42–45
Sustaining effort and persistence
  collective learning, 74
  community goals, 75–76, 75–76 (table)
  foster belonging and community, 78–81
  foster collaboration, 74
  goal setting. *See* Goal setting
  group reflection sheets, 76–77
  group work, 78
  interdependence, 74
  learning goals, 61
  level of effort, 61
  offer action-oriented feedback, 81–84, 82–83 (figure)
  optimize challenge and support, 69–74, 70–73 (figure)
  physical/mental activity, 61

Teaching
  child, 2
  kindergarten teachers, 80
  learning environment, 2, 2 (figure)
  mathematics, 5
  self-advocates, 132
  social and physical environment, 54
  study skills, 198
Temporal contiguity principle, 115
Third-grade classrooms, 36 (table), 51
Thompson, D. R., 141
Thrash, T. M., 274
3-Act Tasks, 51
Treatment Agreement, 109

UDL Math Design Elements, 11, 13 (table)
Unit organizers, 180, 183, 184 (figure)
Universal Design for Learning (UDL), 4, 29, 30
  agency, 14
  characteristics, 11
  classrooms, 10–11
  cultivation condition, 13–14
  design products, 10
  environments, 10
  Guidelines 3.0, 11, 12 (table)
  math proficiency, 14

motivation, 14
outcomes, 11
principles, 11
proactive *v.* reactive, 14
social media platform, 16

Visual vocabulary cards, 145–146, 145 (figure), 146 (figure)
Vocabulary workstation, 149–150

Voice principle, 120
Vygotsky, L. S., 10

Wakefield, D. V., 141
Williams, S., 84
Workstation playlists, 37–38

Zone of Proximal Development (ZPD), 4, 201, 247

Supporting Teachers, Empowering Students

Peter Liljedahl

Fourteen optimal practices for thinking that create an ideal setting for deep mathematics learning to occur.
Grades K–12

Peter Liljedahl, Maegan Giroux, Kyle Webb

Delve deeper into the implementation of the fourteen practices from Building Thinking Classrooms in Mathematics by focusing on the practice through the lens of tasks.
Grades K–5, 6–12

Rachel Lambert

Discover UDL for math, a way to design math classrooms that equips all students for meaningful and joyful math learning.
Grades K–8

Jo Boaler, Cathy Williams

Introduce data science to your students across disciplines with real-world stories and teacher testimonials to transform your classroom experience.
Grades K–8

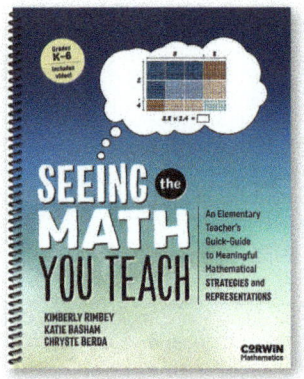

Kimberly Rimbey, Katie Basham, Chryste Berda

Focus on making mathematics meaningful through multiple strategies and representations to help foster a love for mathematics in your students.
Grades K–6

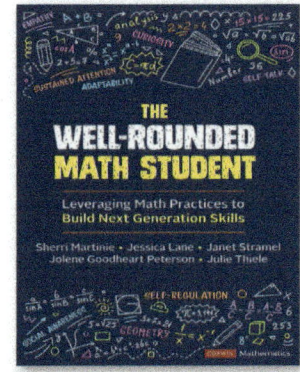

Sherri Martinie, Jessica Lane, Janet Stramel, Jolene Goodheart Peterson, Julie Thiele

Build critical intrapersonal and interpersonal skills *through* mathematics to help all students grow the life-skills they'll carry forever.
Grades K–12

To order your copies, visit corwin.com/math

Our research-based and high-quality content is written by trusted experts and provides clear pathways to helping all students gain access to rigorous mathematics learning; to learn to truly think, reason, collaborate, and fluently discuss mathematics; to form positive and strengths-based mathematical identities; and to see and use mathematics as a tool to effect change in their lives and communities.

Sue Chapman, Holly Burwell, Mary Mitchell

Focus on developing eight essential habits that support mathematical competence and confidence in students through a personalized, practice-based professional learning experience.
Grades K–5

Pamela Weber Harris

Guide students through domains of mathematical reasoning, from counting and adding strategies to more complex proportional and functional reasoning—without resorting to algorithms.
Grades K–12

John J. SanGiovanni, Dennis McDonald

Enhance your students' understanding and engagement in geometry, measurement, and data while also fostering a deeper connection between math and the real world.
Grades K–5

Vanessa "The Math Guru" Vakharia

Equip students to develop the skills they need to truly believe anything is possible, even a better relationship with math!
Grades K–12

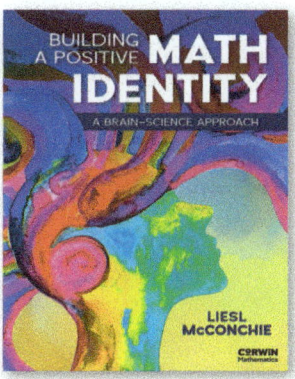

Liesl McConchie

Reexamine what it means to have a positive math identity—and learn to use brain-based tools in a humorous and friendly way to build on a positive math identity for your students.
Grades K–12

Irina Lyublinskaya, Xiaoxue Du

Integrate AI literacy into K–12 classrooms, blending theory, practical lesson plans, and ethical considerations to empower students as critical thinkers.
Grades K–12

To order your copies, visit corwin.com/math

CORWIN Mathematics

CORWIN

To help every educator help every student

We believe that every single student deserves a great education

We believe that knowing our impact is both a privilege and a responsibility

We believe that a fair, stable, and thriving society is built on education